METAPHYSICS AND EXPLANATION

METAPHYSICS
AND
EXPLANATION

Proceedings
of the
1964 Oberlin Colloquium
in Philosophy

Edited
by
W. H. Capitan
and
D. D. Merrill

UNIVERSITY OF PITTSBURGH PRESS

by C. Tinling & Co. Ltd., Liverpool, London and Prescot

PREFACE

The essays in this volume were presented at the fifth annual Oberlin Colloquium held on April 10-12, 1964, by the Oberlin College Department of Philosophy. As in previous Colloquia, philosophers were invited to present papers on topics of their own choice for intensive discussion by commentators and by other philosophers invited to attend. The resulting essays, all of which are published here for the first time, show the influence of analytic philosophy while reflecting some of its diverse forms.

In the first essay, Stephan Körner disputes two fundamental parts of the standard "deductivist" account of scientific testability. He asserts, first, that the testable consequences of a scientific theory are not empirical propositions, but rather idealized theoretical counterparts of empirical propositions. He also denies that the reasoning from premises to conclusions of scientific theories is always deductive, using statistical theory and quantum mechanics as counterexamples.

In the next essay, J. J. C. Smart distinguishes and relates two senses of 'nonsense': first, 'nonsense' used in the sense of "meaningless"; second, "nonsense" used in the sense of its being "without sense" to assert an egregious falsehood. He argues that these two senses of the word "nonsense" are closely related and that there is no sharp distinction between science as concerned with truth and philosophy as concerned with concepts. Philosophers need not become amateur scientists, however, since they, as nonspecialists, have an important role—to seek in the light of modern science the most probable answers to questions whose answers require an integrated conceptual scheme.

In the first symposium, Joel Feinberg argues it is consistent with common sense to speak of the causes of voluntary actions, contrary to Hart and Honoré's view in *Causation and the Law*. There are many more ways of causing someone to act than by forcing or inducing him. Donnellan agrees but suggests that there are significant differences in our treatment of these causes and those of natural events. Besides claiming Feinberg in part misinterprets Hart and Honoré, Lehrer argues that even if it is consistent to say a voluntary action is caused, there is still reason to doubt that determinism is consistent with common sense beliefs about voluntary actions. In reply, Feinberg grants Donnellan that there are significant differences, but adds that questions of a form readily applicable to mere events are also applicable to voluntary actions. Feinberg also tries to remedy the misinterpretation cited by Lehrer, but denies that the misinterpretation lessens the force of his counterexamples. He finds Lehrer's second charge irrelevant to his thesis.

In the second symposium, Nicholas Rescher distinguishes between two kinds of metaphysical tasks: *taxonomic* metaphysics, concerned with delimiting the various sorts of things, and *evaluative* metaphysics, concerned with evaluating the various sorts of things. He argues that the latter is a legitimate enterprise, that it may be looked at theologically through the quasi-ethical evaluation of hypothetical acts of creation, and that it is implicit in ordinary ethical evaluations. Lewis White Beck and Thomas Patton find Rescher's metaphysical values to be unnecessary additions to ordinary ethical values, a claim which Rescher denies in his final comments.

Finally, Herbert Hochberg discusses alternative analyses of things and their qualities. He attempts to show that the usual arguments in support of bare particulars are inadequate, and that things can be analyzed as combinations of qualities without encountering the usual problems of identity and difference. He also discusses the view that things are combinations of instances of qualities. Richard Severens also rejects bare particulars, but finds that the combination-of-qualities view violates the restrictions of type theory. J. M. Shorter argues that all three analyses are either unintelligible or else compatible with one another. In reply, Hochberg finds Severens' arguments against bare particulars unconvincing and finds no violations of type theory in the combination-of-qualities view. He argues that Shorter's objections reflect the view that ontological analyses are simply esoteric ways of talking about familiar things.

The contributors to this volume represent ten colleges and universities in this country and abroad: Mr. Körner from Bristol, Mr. Smart from Adelaide, Mr. Feinberg from Princeton, Mr. Donnellan from Cornell, Mr. Lehrer and Mr. Beck from Rochester, Mr. Rescher from Pittsburgh, Mr. Patton from Pennsylvania, Mr. Hochberg from Indiana, Mr. Severens from Ohio State University, and Mr. Shorter from the University of Canterbury, Christchurch, New Zealand.

Members of the Oberlin College Department of Philosophy wish to thank all of the participants for helping to make the 1964 Colloquium a profitable and stimulating experience. They also wish to thank Oberlin College for its continuing support of the Colloquium and for its assistance in the publication of this volume. Finally, the editors are grateful to the contributors to this volume and to the staff of the University of Pittsburgh Press for their generous co-operation and assistance in preparing this volume.

CONTENTS

PREFACE 5

ON DEDUCTIVISM AS A PHILOSOPHY OF SCIENCE . . 9
Stephan Körner

NONSENSE 20
J. J. C. Smart

Symposium

CAUSING VOLUNTARY ACTIONS 29
Joel Feinberg

COMMENTS 48
Keith S. Donnellan

COMMENTS 52
Keith Lehrer

REJOINDERS 55
Joel Feinberg

Symposium

EVALUATIVE METAPHYSICS 62
Nicholas Rescher

COMMENTS 73
Lewis White Beck

COMMENTS 76
Thomas E. Patton

REJOINDERS 79
   Nicholas Rescher

Symposium

THINGS AND QUALITIES 82
   Herbert Hochberg

COMMENTS 98
   Richard Severens

COMMENTS 101
   J. M. Shorter

REJOINDERS 106
   Herbert Hochberg

ON DEDUCTIVISM AS A PHILOSOPHY OF SCIENCE

STEPHAN KÖRNER

It is a truism that scientific hypotheses and theories are tested by comparing some of their consequences with experience. From it must be distinguished a philosophical position sometimes called "deductivism." Its main theses are (1) that the testable consequences of a scientific theory are empirical propositions (and not, for example merely "symbolic representations" of experience in Duhem's sense of the term); (2) that within a scientific theory the reasoning from premisses to conclusions is always deductive; (3) that there is no logic of induction, and that induction cannot be justified. Whereas the third of these theses is still highly controversial, the first and second are, it seems, upheld by most contemporary philosophers of science, which makes their examination all the more desirable. In this paper I propose to show that both of them must be rejected.

As is customary, I shall distinguish between the logico-mathematical or formal concepts and propositions of a scientific theory on the one hand and its substantive concepts and propositions on the other. The logic in question is the two-valued propositional logic, quantification theory or predicate-logic of first order, and the theory of equality. I shall regard as mathematics any extension of this logic into a predicate-logic of higher order or a theory of sets, sufficient to accommodate the mathematical content and techniques employed by the scientific theory under consideration. As to where the line should be drawn, in a scientific theory, between the logico-mathematical framework and the substantive content of a scientific theory, there is less agreement. Thus, partly depending on their different notions of probability, some writers, e.g. Carnap, regard probability-theory as logico-mathematical, while others, e.g. von Mises, regard it as a substantive theory. Such delimitation disputes will, however, not affect the subsequent argument.

1. *Examination of the first thesis.* While both the formal framework and the substantive content of a scientific theory, e.g. physics, change in the course of time, the substantive content changes the more quickly of the two. For this and other reasons, the systematic unity of the formal and the substantive part of a theory may be more or less perfect. It is perfect if the two parts together constitute a hypothetico-deductive, or axiomatic, system. In this case, and in this case only, is all the reasoning within the theory deductive, consisting in the deduction—by means of the logico-mathematical

postulates and rules of inference—of substantive theorems from the conjunction of the theory's substantive postulates; or of a state-description from the conjunction of the substantive postulates together with another state-description. Considering a theory T_0 and writing 'a_0' for the conjunction of its substantive postulates, 't_0' for one of its substantive theorems, 'b_1' and 'b_2' for two state-descriptions, and '$\vdash_L$' for deducibility by means of the logico-mathematical framework, we may represent the two kinds of deduction schematically by

$$a_0 \vdash_L t_0 \qquad (1)$$
and
$$(b_1 \wedge a_0) \vdash_L b_2 \qquad (2)$$

For example, if a_0 is the conjunction of the substantive postulates of classical mechanics, then t_0 is a theorem of this science; and if b_1 is a state-description of a mechanical system in terms of positions and momenta at a certain time, b_2 is a state-description of this system at some other time.

The first thesis of deductivism implies that b_1 and b_2 are empirical propositions. The reason why this thesis must be rejected lies in the restrictions which the logico-mathematical framework L of T_0 imposes on any concepts and propositions capable of being accommodated into L and so into T_0. In consequence of these restrictions, empirical concepts such as those of empirical (operationally ascertainable) length, weight, duration, etc., have to be modified into or, what comes to the same, replaced by, other concepts, such as can be fitted into T_0 and can occur as constituents of b_1 and b_2, formulated in terms of the vocabulary of T_0. The most general restrictions, which are imposed upon any theory by its logico-mathematical framework, are due to the structure of classical logic alone. Since I have discussed them elsewhere in some detail,[1] a few brief remarks will be sufficient here.

As a consequence of the role of definition by examples and counterexamples of their applicability, many empirical predicates are inexact, i.e., admit of borderline or neutral cases to which they or their complements are applicable with equal correctness. This means that in a logical theory which does justice to empirical discourse, one must admit (if o is an empirical individual and P an empirical predicate), not only the possibilities $P(o)$—'$P(o)$' is true—and $\neg P(o)$—'$P(o)$' is false—but also a third possibility, say, $*P(o)$—'$P(o)$' is neutral—which is provisional in the sense that one is, by the rules governing P, free to change it either into $P(o)$ or into $\neg P(o)$. A three-valued propositional logic and quantification theory has been constructed by Kleene[2] in which the third truth-

---

[1] In "Deductive Unification and Idealisation," *British Journal of the Philosophy of Science*, XIV, 1964, pp. 274–289.
[2] *Introduction to Metamathematics*, New York, 1952, Sec. 64.

value is provisional and not final as the other two. However, whereas in Kleene's system it may be that the third truth-value cannot be turned into one of the other two, such a change is always possible for borderline cases and neutral propositions. By taking this into account one can develop a logical system which does justice to inexact predicates, i.e., predicates admitting of neutral cases, and to neutral propositions, i.e., applications of inexact predicates to their borderline cases. Now it is clear, or can by a more detailed discussion be made clear, that if the classical logic is to become available as an instrument of deduction yielding substantive conclusions from substantive premises, all inexact empirical predicates and neutral propositions must be first of all replaced by exact predicates and definite propositions.

If the logic underlying a scientific theory comprises also the logic of equality, then in addition to the elimination of inexactness, a further modification of the field of empirical discourse is made necessary if it is to be deductively unified by classical logic. This, as Poincaré and many others after him have pointed out, is the replacement of nontransitive empirical equality by transitive mathematical equality: from the empirical indistinguishability in a certain respect of a pair of empirical objects a and b and a pair b and c, the indistinguishability in this respect of the pair a and c does not follow, whereas mathematical equality is, of course, transitive. Since the notion of transitive equality is involved in all scientific reasoning about the results of measurement, the modification affects all quantitative reasoning.

Quantitative reasoning in the full sense of the term imposes, of course, further restraints on the scientific theory within which it takes place. The nature of the concepts of a quantity, of the addition of quantities, of the multiplication of a quantity with a number, of the definition of units, etc., has been investigated by Helmholtz and others. A clear restatement of the theoretical conditions of measurement has recently been given in various works by K. Menger, who in addition provides an acute analysis of functional relations as used in scientific theories.[3] These conditions demand an isomorphism between physical and mathematical addition and other physical and mathematical operations, which does not hold between them. In order to satisfy this demand, one has to modify the notions of empirical quantities and the operations upon them.

As the logico-mathematical framework is extended, still further modifications become necessary in the field of empirical discourse which is to be fitted into it. I may mention, for example, the

---

[3] See, e.g., his "Mensuration and Other Mathematical Connections of Observable Materials" in *Measurement: Definitions and Theories* ed. by C. W. Churchman and P. Ratoosh, New York, 1959.

replacement of the empirical notion of continuity by a mathematical notion defined as a type of order in a nondenumerable totality of elements.[4] The deductive unification of experience by classical logic and its extensions is thus *ipso facto* an idealization.

In the case of many scientific theories, idealization by deductive unification is supplemented by another type of idealization, which might be called "deductive abstraction." It will suffice to give an example. Let us assume that a material object is observed with a view to formulating a state-description of classical mechanics. The thing possesses a variety of perceptual properties and stands in perceptual relations to other things. These perceptual characteristics are either inexact or in case they are determinables, such as color, extension, weight, position, etc., exact. Moreover, of the thing's perceptual characteristics not all are relevant to classical mechanics. Some are clearly relevant; more particularly relevant are perceptual weight, for which I shall write 'W_o,' and relative position (with respect to some fixed object), for which I shall write 'Q_o.' The other perceptual characteristics of the thing, which I shall assume to constitute the range of the predicate variable 'P,' are irrelevant.

Now whatever possesses perceptual weight or perceptual position or both, necessarily possesses some other perceptual characteristic different from either of them. This may be schematically expressed by:

(Aa) $\vdash (x) [(x \varepsilon W_o) \rightarrow (\exists P)(x \varepsilon P)]$ and
(Ab) $\vdash (x) [(x \varepsilon Q_o) \rightarrow (\exists P)(x \varepsilon P)]$,

where '$\vdash$' expresses deducibility between propositions with exact predicates as constituents, which is permissible here since W_o, Q_o and the values for the variable P are determinables and thus exact. On the other hand the individuals of classical mechanics, its particles, possess *only* weight (defined in terms of mass and acceleration due to gravity) and position. It is not simply the case that the presence of other characteristics is ignored: their absence is postulated. Thus heat could not be conceived as the energy of a system of moving particles, unless the particles were assumed to be neither hot nor cold.

That a particle of classical mechanics has weight and position *only* may be schematically expressed by:

(Ba) $\vdash (x) [(x \varepsilon W_o') \rightarrow \rightarrow (\exists P)(x \varepsilon P)]$ and
(Bb) $\vdash (x) [(x \varepsilon Q_o') \rightarrow \rightarrow (\exists P)(x \varepsilon P)]$.

I have written 'W_o'' here instead of 'W_o' as also 'Q_o'' instead of 'Q_o' since propositions A and B determine different predicates of

---

[4] For an analysis of the concept of empirical continuity, see my article in *Monist*, XLVII, 1962, pp. 1–19.

weight and position, namely perceptual on the one hand and "mechanical" on the other.[5] A comparison of A and B shows moreover their incompatibility, and that the extensions of W_o (Q_o) and W'_o (Q'_o) and of all their respective subclasses are mutually exclusive, i.e., schematically, that

(C) $\vdash(x) [(x \varepsilon W_o) \rightarrow \rightarrow (x \varepsilon W_o')]$.

In general, when deductive abstraction is applied to perceptual characteristics, it consists in distinguishing between relevant perceptual determinables and, so to speak, in removing the deducibility-relation which holds between any relevant and any irrelevant determinable. It results in the formation of new "abstract" determinables, every abstract determinable and its perceptual counterpart being mutually exclusive. Abstraction thus constitutes a further method of idealization, independent of that imposed upon a scientific theory by its logico-mathematical framework. Whereas the latter draws sharp demarcation-lines through indefinite conceptual borders, the former cuts away, as it were, what from the point of view of a theory is irrelevant.

Returning now to the first thesis of deductivism, it has become clear that in the schema

$$(b_1 \wedge a_o) \vdash_L b_2 \qquad (2)$$

b_1 and b_2 are *not* empirical propositions, since all their constituent concepts have been replaced in one or more of the ways just described by counterparts which are no longer descriptive, but idealized characteristics, of experience. The state-descriptions b_1 and b_2 must, therefore, be identified with empirical propositions, say e_1 and e_2, respectively. The schematic representation of testable scientific reasoning would consequently have to be represented by a more complex schema which, apart from the theoretical derivation (2), also takes into account the identifications of empirical with idealized predicates, propositions and individuals, and the context in which these identifications are assumed to be permissible. This more complex schema might be formulated, say as

$$[e_1 \wedge c_o \wedge (e_1 \approx b_1) \wedge ((b_1 \wedge a_o) \vdash_L b_2) \wedge (b_2 \approx e_2)] \rightarrow e_2 \qquad (3)$$

Here e_1 and e_2 are empirical, b_1 and b_2 are theoretical propositions, i.e., are formulated in terms of predicates restricted by the logico-mathematical framework of the theory in question and (usually) by deductive abstraction. c_o states the context of the identification, the

---

[5] Strictly speaking, one should, of course, compare perceptual mass (as determined, for example, by weighing) and perceptual (spatio-temporal) position on the one hand, with mass and (spatio-temporal) position, as defined in classical mechanics, on the other.

arrow '→' expresses the 'if—then' and '∧' the conjunction sign of the logic of empirical discourse, which admits both exact and inexact predicates.

Turning now to the skeleton proposition (3), instead of attempting to clothe it any further I propose to consider two opposite strategies by which a deductivist might attack it. He might argue that in proposition (3) everything but the theoretical derivation either is redundant, or else can be incorporated into the theoretical derivation.

The redundancy argument may be put thus: once the identification of an empirical with a theoretical proposition, e.g., $e_1 \approx b_1$, has been justified, the empirical proposition—and with it all the components of (3) involving it—can be discarded in favor of the theoretical proposition. But this is not so, at least not if the empirical proposition describes an experiment or an observation sufficiently important for a record of it to be preserved. Any such observation or experiment has to be described in empirical and not in theoretical terms. The proposition describing it must be about things (processes or other empirical individuals), e.g., billiard balls and lamps, as opposed to particles and light rays, and about actual operations performed upon them, i.e., empirical operations and not thought-experiments. Experiments and observations, as a glance at any textbook will show, are always expressed by empirical propositions.

This does not mean that these propositions and their constituents do not include reference to theories. For example, the predicate 'measuring instrument' or, more particularly, 'interferometer' is empirical and describes things which can be bought in shops, even though their use and, to some extent their manufacture, presuppose a knowledge of a theory. It is one thing to distinguish between empirical and theoretical propositions and quite another to deny that emprical propositions and predicates may not have theoretical propositions and predicates as constituents. Indeed proposition (3) is an example of such an empirical proposition; and logical abstraction (i.e., the formation of open propositions and class-abstracts from it) would yield examples of empirical predicates with theoretical constituents.

Apart from the fact that empirical propositions, even after their identification with theoretical propositions, are needed as records of experiments and observations, they are also indispensable for scientific co-operation. An empirical proposition which is by one group of scientists identified with a theoretical proposition belonging to one theory, must remain available in order to be identified by another group of scientists with a theoretical proposition belonging to an altogether different theory. Again scientists often depend on the help of technicians and even amateurs, who are not capable of, or interested in, the identification of empirical propositions with the theoretical propositions of a specific theory.

So much for the failure to save the first thesis of deductivism by reducing proposition (3) to proposition (2), by declaring the other components of (3) as redundant. I now turn to the opposite defense-strategy, which is to argue that any scientific theory is rich enough in its means of expression and inference to allow for the deduction of an (unmodified) empirical proposition, say e_2, from a conjunction of another such empirical proposition and the substantive postulates of the theory. The aim of this strategy is to replace proposition (3), not by

$$(b_1 \wedge a_0) \vdash_L b_2 \qquad (2)$$

but by

$$(e_1 \wedge a_0) \vdash_L e_2 \qquad (4)$$

But this is clearly hopeless since the constituent predicates of e_1 and e_2 are not admitted into the theory, in virtue of the restrictions imposed on all its predicates by the classical logic of propositions and quantification theory, by the logic of equality and—in the case of many theories—by the conditions of measurement, by the applicability of the classical mathematics of real numbers and by deductive abstraction.

Conflation of propositions (4) and (3) is the reason for an unrealistic account of the manner in which conflicts between testable reasoning and actual tests are resolved. According to this account, if e_1 is true and e_2 turns out to be false, then the theory, or the conjunction of its substantive postulates a_0, is falsified. In fact, however, a conflict between testable reasoning and actual tests may be traced to shortcomings either in the substantive postulates or in the conditions—expressed in (3) by c_0—for identifying the empirical with the theoretical propositions. Thus the experimental set-up may have failed to exclude theoretically irrelevant, disturbing factors, known or unknown. If, for example, in a mechanical experiment $\neg e_2$ is due to the fact that the bodies on which it was performed are electrically charged, the mechanical theory is not falsified. Moreover, if e_2 turns out to be false, one may, even if one cannot discern a disturbing factor, still decide to search for one, rather than regard the theory as falsified.

Proposition (3), but not the inadequate (4), allows for this choice between the rejection of the theory and the search for external disturbing factors, i.e., for conditions which the identifications have to satisfy. The history of science is full of examples of such kinds of choice, and it does not support the view that every conflict between testable reasoning and actual tests amounts to a falsification of the

16   ON DEDUCTIVISM AS A PHILOSOPHY OF SCIENCE

theory employed in the reasoning. T. S. Kuhn[6] has shown convincingly that the development of scientific theories does not bear out the identification of "anomalous" with falsifying experiences "because," to quote his words, "it is just the incompleteness and imperfection of data-theory fit that, at any time, define many of the puzzles that characterize normal science."

2. *Examination of the second thesis*. In criticizing the first thesis of deductivism I have assumed that the systematic unity of a scientific theory is perfect, in the sense that its logico-mathematical framework and its substantive postulates together constitute a deductive or a axiomatic theory, in which unbroken deductive chains lead from substantive postulates to substantive theorems, as schematically expressed in (1), or from state-descriptions in conjunction with the substantive postulates, to different state-descriptions, as expressed schematically in (2). I shall now argue that this assumption is not in general true and, hence, that the second thesis of deductivism—that the reasoning from premisses to conclusions within a scientific theory is always deductive—must be rejected. To convince ourselves of this, it will be sufficient to show that two important scientific theories actually exhibit deductive disconnections in their internal chains of reasoning. The arguments to this effect will be similar to those by which I have tried to show that a theoretical predicate produced by deductive abstraction from an empirical determinable is incompatible with it. The difference is merely that, in the cases to be discussed, both the predicate subject to deductive abstraction and the predicate resulting therefrom are theoretical.

My first example is statistical theory. This can be, and has been regarded, either as an independent scientific theory or as part of the logico-mathematical framework of other theories, e.g., quantum mechanics and genetics. The empirical subject matter of statistical theory is mass phenomena, in particular empirical proportions, sequences of empirical proportions, the degree of stability of such sequences, etc. In considering, let us say, the proportion of Q's in a class of P's, where P and Q are empirical predicates, let us write

$$\phi_n = \frac{n(P \wedge Q)}{n(P)}$$

for 'the proportion of Q's in a class of nP's'; and '$\{\phi_n\}$' for a sequence starting with ϕ_1 and ending with ϕ_n, where n is a natural number. By incorporating statements involving 'P,' 'Q,' 'ϕ_n' and '$\{\phi_n\}$' into classical logic one imposes upon them the restrictions of this logic. This, as has been shown, amounts to eliminating inexactness and to replacing nontransitive empirical by transitive mathematical

[6] *The Structure of Scientific Revolutions*, Chicago, 1962, pp. 145f.

equalities. A further modification involved in the statistical characterization of mass-phenomena consists in assuming that n although finite, may be as large as needed. In order to indicate the transition from the empirical description of mass-phenomena to their *finite* idealization different symbols should be used. I shall, therefore, indicate the *finite* idealization of '$\{\phi_n\}$' by '$\{f_n\}$.'

Now statistical theory is to a large extent concerned with (1) the structure of finite sequences $\{f_n\}$ and, more generally, with so-called sampling distributions; (2) the structure of infinite sequences, say $\{F_n\}$ (the limits of which, as n tends to infinity, are defined as their probabilities) and, more generally, with so-called probability-distributions; and (3) the conditions under which a sampling distribution can be identified with a corresponding probability-distribution—the conditions being known as significance-tests. That the reasoning is not deductive, which leads from a probability-distribution *plus* the result of certain significance-tests to a frequency-distribution, or in the opposite direction, is pretty obvious, and is clearly explained in the textbooks of statistical theory.[7] Indeed the "identification" of a finite with an infinite sequence is the identification of two incompatible concepts and is more accurately described as a gap in a deductive chain, brought about by replacing one concept by another.

Braithwaite,[8] approaching the subject from a different angle, also reaches the conclusion that statistical theory is not a hypothetico-deductive system in the customary sense; and he so reconstructs it, that what seems to be its "lowest level hypotheses" (e.g., that 51% of children born are boys) require to be treated as "highest-level hypotheses in an unendingly descending hierarchy, the deductive principles used in this deduction being those of a special branch of mathematics (to be called *class-ratio* arithmetic)." But in the present context it makes little difference whether the gap in question cannot be deductively bridged at all or only in an infinite number of steps.

My second example is quantum mechanics. In its orthodox versions quantum mechanics is not a "generalization" of classical mechanics; and classical mechanics itself cannot be simply, if at all, regarded as a "special" or "limiting" case of quantum mechanics, which arises in situations in which the quantum of action can be set equal to zero. That the relation between the two theories is more complex than this is, in one way or another, recognized by most physicists.

Classical mechanics is part of quantum mechanics in the sense that all quantum mechanical measurements are made by means of

---

[7] See, e.g., H. Cramér, *Mathematical Methods of Statistics*, Princeton, 1946, pp. 334f.
[8] *Scientific Explanation*, Cambridge, 1953, e.g., p. 118.

classical apparatus, that is to say in terms of procedures which in the first instance are conceived as subject to the laws of classical mechanics. To take a crucial example, determination of the classical momentum and position of an object is based on the assumption that the results of these measurements are independent of their order or, in more technical language, that momentum and position are commutative quantities. Writing '$x \, \varepsilon \, M_c$' for 'x has classical momentum' and '$x \, \varepsilon \, O$' for 'the measurements of x are independent of their order,' we have

(D) $\vdash(x) \, [(x \, \varepsilon \, M_c) \rightarrow (x \, \varepsilon \, O)].$

In quantum mechanics, on the other hand, momentum and position of an object—and all other so-called conjugate quantities—are conceived as non-commutative, their non-commutativity being either among the substantive postulates of the theory or among its substantive theorems. Writing '$(x \, \varepsilon \, M_q)$' for '$x$ has quantum-mechanical momentum,' we have, therefore,

(E) $\vdash(x) \, [(x \, \varepsilon \, M_q) \rightarrow \rightarrow (x \, \varepsilon \, O)].$

A comparison of (D) and (E) shows that M_q and M_c are mutually exclusive. The statement that M_q is measured classically means that what is measured is M_c, which is then identified with (or, better, replaced by) M_q.

Just as no deductive sequence of classical physics ends in an empirical proposition, and just as this deductive gap is closed by the identification of a theoretical proposition of classical physics with an empirical proposition, so no deductive sequence of quantum mechanics ends in a proposition of classical physics, and the deductive gap is closed by "identifying" a theoretical proposition of classical with a theoretical proposition of quantum mechanics. Since many theoretical structures, in particular statistical theory and quantum mechanics, are unhesitatingly considered as scientific theories, it has to be admitted that not all scientific theories are hypothetico-deductive. In order to emphasize the bridging of deductive gaps by identifications of the kind described, the symbol '$\vdash_L$' in proposition (3) should be dropped in favor of, say, '$\vdash_{L,I}$' which refers to both the deductive and the identificatory steps in the reasoning.

In conclusion I set down two theses by which, as a result of the preceding examination, I should propose to replace the first and second theses of deductivism: (1) The testable consequences of a scientific theory are not empirical, but are idealized theoretical propositions; the reasoning by which they, and the theory containing them, is tested conforms to proposition (3)—with '$\vdash_{L,I}$' substituted for '$\vdash_L$'—and involves the identification, in specified contexts, of these theoretical with corresponding empirical propositions. (2) The reasoning from the substantive premisses to the substantive conclusions of scientific theories proceeds not only by deduction, but also by the identification of deductively unconnected theoretical predicates and propositions.

NONSENSE

J. J. C. SMART

Should we say that the sentence 'Prime numbers are married' expresses a falsehood or should we say that it is nonsense? Here 'nonsense' is used in the sense of 'meaningless,' i.e., in one sense of the phrase 'without sense.' Now just as in another sense of the words 'without sense' it would be without sense to go seeking kangaroos in Manhattan, so in this last sense of 'without sense' it would be without sense to assert an egregious falsehood. Often therefore the word 'nonsense' is used to express vehemently the thought that someone has uttered an egregious falsehood, and often the vehemence is all the greater when it is hard to be sure, or to prove rationally, that the alleged egregious falsehood really is a falsehood. For example, a man who is very prejudiced in favor of capital punishment is likely to answer with "Nonsense!" any suggestion, however reasonable, that a criminal should be reprieved.

We may therefore distinguish two senses of 'nonsense' which answer roughly to those of 'meaningless' and 'false,' respectively. Nevertheless, I shall suggest in this paper that these two senses of the word 'nonsense' are more closely connected than is sometimes thought, and that often it may be a matter for choice which should be employed in a given case. Consider the sentence 'This electron has both a definite position and a definite momentum.' In accordance with modern quantum mechanics we should wish to reject this sentence. But on what grounds should we do so? Is it to be rejected because it is meaningless or is it to be rejected because it expresses a falsehood? I doubt whether a physicist would normally be much interested if we put this question to him. He does not want us to use this sentence, and that is the main thing: whether it should be stigmatized as false or meaningless is a secondary matter.

For example, we could take the line that in the language of modern physics the sentence in question is somehow a wrongly constructed or "ungrammatical" one. But equally we might take the line that it must be rejected because, although it is a correctly formed sentence, it leads to contradiction when it is conjoined to the principles of quantum mechanics. Since quantum mechanics has never been exhibited as a fully formalized system, we cannot decide between these alternatives. It is possible that there could be two equally good formalizations of quantum mechanics: (1) a formalization whereby the formation rules of the system do not enable us to construct a sentence asserting the simultaneous definite position and definite momentum of an electron; (2) a formalization according to which a sentence asserting the simultaneous definite position and definite momentum of an electron would be correctly formed but inconsistent with the axioms. In fact, I suspect that the second of these two alternatives would prove to be the simpler one. If this is so, then it would be most natural to suppose that a physicist who said that it was nonsense to assert the simultaneous position and momentum of an electron would be using the word 'nonsense' in the sense of 'egregious falsehood,' not in the sense of 'meaningless.'

It would seem, therefore, that in science there is not a sharp line between the separation of the meaningful from the meaningless and the separation of truth from falsehood. Often the same thing could be looked at in either way. A corollary perhaps is this: that there is not such a clear separation as is often thought between philosophy conceived as the study of how not to talk nonsense and philosophy conceived as the assertion of truths about the world.

In some of his classes at Oxford after the war, Gilbert Ryle used illuminatingly to retail some of the history of the preoccupations of philosophers with the nature of their own subject.[1] In the last century, philosophy used to be thought of as "mental science." It did not appear to be the science of the material world because this was clearly the subject matter of physics, chemistry, and biology. But it still seemed plausible to suppose that philosophy was concerned with the mind. The idea that philosophy is mental science still survives fossilized in the title of the English philosophical journal of which Ryle himself is now the editor. This confusion of philosophy with psychology was facilitated by the pre-scientific character of psychology at the time and also by psychologistic misconceptions about the nature of logic. (Consider the title of Boole's book, *The Laws of Thought*.)

When around the turn of the century it began to appear that psychology was one of the empirical sciences, the worry of philosophers about the legitimacy of their subject could be allayed by the Platonizing phase of philosophy, as in Russell's early writings. It could be thought that philosophy differed from the sciences in that it was the study of eternal objects, such as universals and propositions. Russell's Platonism was quite different from the more recent sort we associate with Quine. According to the early Russell, universals were known by some sort of intellectual intuition, whereas for Quine his Platonic objects (classes) are theoretical posits just as electrons are. We need mathematics in order to do physics, and since mathematics needs classes, so does physics.

[1] For a concise account of some of these matters see G. Ryle, "Ludwig Wittgenstein," *Analysis*, XII, 1951-52, pp. 1-9.

When the sort of Platonism which we get in the early Russell went out of fashion (partly, I think, because the intellectual intuitions required for it were hard to reconcile with a scientifically acceptable account of cognition), then, once more, there arose the question of the demarcation between philosophy on the one hand and the special sciences on the other hand. According to Ryle, the answer given by Platonism had the virtue of being correct negatively: it was right in denying that philosophy was about either the mental or the physical world. On the positive side, Platonism was wrong: Ryle's answer was that philosophy was "talk about talk," although not in the way in which ordinary grammar and philology are. Philosophy, on this view, is concerned with the conceptual structure both of the special sciences, and of our common sense talk, including our moral, legal, aesthetic talk, etc. Philosophy came to be thought of as a second-level subject which dealt with the conceptual confusions which we can get into if we mishandle our first-level talk. Such confusions are especially likely to occur where different sorts of talk impinge on one another, for example on the borderline between physics and biology, or between psychology and jurisprudence.

If philosophy was concerned with conceptual confusions, then it was concerned specifically with the distinction between the meaningful and the meaningless. It is not, of course, concerned with meaninglessness in the sense in which "Jabberwocky" or a verse by Edward Lear is meaningless, but in the way in which Lewis Carroll's description of a grin without a cat is meaningless.

The view that philosophy is concerned with sense and nonsense has been seen by Ryle[2] (approvingly) and by Popper[3] (disapprovingly) as a generalization of Russell's theory of types. Indeed they both interpret Wittgenstein as having propounded such a generalization in the *Tractatus*. When Russell put forward his famous paradox of the class of all classes not members of themselves, he brought to the attention of philosophers the idea that meaninglessness is insidious, interesting, and important. Russell, with his theory of types, suggested that both the sentence 'The class of all classes not members of themselves is a member of itself' and the sentence 'The class of all classes not members of themselves is not a member of itself' are meaningless. It therefore appeared that there are rules of sentence construction, the violation of which can render a sentence meaningless, and yet which are far from obvious and which are not encompassed by ordinary grammar.

---

[2] *Op. cit.*

[3] K. R. Popper, "The Demarcation between Science and Metaphysics," in *The Philosophy of Rudolf Carnap*, ed. by P. A. Schilpp, La Salle, Illinois, 1963. This article has also been published in Popper's *Conjectures and Refutations*, London, 1962.

Although Russell seemed to regard this sort of danger of meaninglessness as localized and sporadic, the generalization of the idea led to the view that this subtle sort of meaninglessness might break out all over the place. It began to be suspected that this sort of meaninglessness characterized the peculiar utterances of the metaphysicians. Philosophy began to be seen as a technique for exposing the subtle traps which language sets, and which may lead to nonsense in all realms of discourse, not just in the special field of set theory.

Popper has pointed out[4] that if this is the analogy which lies behind the Wittgensteinian and neo-Wittgensteinian conceptions of philosophy, then it has a very weak basis. If we look at the sentences about the class of all classes not members of themselves from Russell's point of view, then they do seem to be meaningless, for they violate Russell's type rules. From Russell's point of view they are ungrammatical or wrongly formed sentences, just as 'Runs is or' is, although less obviously so. But we can just as well look at Russell's paradox from the point of view of other set theories, in which the sentence 'The class of all classes not members of themselves is a member of itself' is not meaningless but expresses a falsehood. If there is no class of all classes not members of themselves, it follows that the sentence 'The class of all classes not members of themselves is a member of itself' and the sentence 'The class of all classes not members of themselves is not a member of itself' must both be false if they are meaningful. Something like this is the way taken both by Zermelo's set theory and by Quine's *Mathematical Logic* (ML). In Zermelo's theory the objectionable sentences cannot be proved because there is no way of forming the class of all classes not members of themselves. We could take Russell's paradox as a way of disproving the existence of this class.[5] Similarly, in ML, Russell's paradox shows that the class of all classes not members of themselves does not exist. However, the class of all *sets* that are not members of themselves does exist.[6] (Sets are classes capable of being members of other classes, whereas in ML there can be ultimate classes—classes incapable of membership.)

It follows that if the Wittgensteinian or neo-Wittgensteinian concept of philosophy is an extension of Russell's theory of types, then it is a very questionable one. Popper is surely correct in holding that the possibility of alternative ways of dealing with Russell's paradox shows the matter in a different light. The possibility of

---

[4] *Op. cit.* See p. 194 in the Carnap volume or p. 263 in *Conjectures and Refutations*.
[5] See L. Goddard, "Sense and Nonsense," *Mind*, LXXIII, 1964, pp. 309-331, especially pp. 324 and 326.
[6] See W. V. Quine, *Set Theory and Its Logic*, Cambridge, Massachusetts, 1963, p. 309.

alternative ways of dealing with Russell's paradox has as its analogue alternative ways of dealing with certain controversial sentences: it may be possible to develop science or metaphysics in such a way that they become meaningless, but equally they may find their place as truths or falsehoods.

A tolerant attitude to different set theories should therefore generalize into a tolerant attitude toward the conception of philosophy as concerned with truth and falsehood, rather than merely with meaningfulness and meaninglessness. The generalization of Russell's theory of types suggested that there was a kind of grammar, "logical grammar," which might be thought of as ordinary grammar done more rigorously and with a finer accuracy, and with a disregard for the accidental philological details, such as irregular verbs, with which ordinary grammar also concerns itself. Thus it might be thought that just as at school we learn to distinguish crudely between nouns, verbs, adjectives, adverbs, and so on, when we do philosophy we learn to make finer distinctions, such as Ryle's distinction between task words and achievement words. Again, following Michael Shorter, in his article "Meaning and Grammar,"[7] we may assimilate the incorrectness of 'Virtue is square' to that of 'Virtue is but.' After all, is ordinary grammar to be fossilized at a particular stage? Is it not arbitrary to say that 'Virtue is square' is grammatical, while 'Virtue is but' is ungrammatical? Are they not *both* deviant sentences, even though one is legalized by school grammar while the other is not? Such, in outline, is the sort of case that can be made for "logical grammar": its weakness is that grammar may be concerned with the rules which correspond to a certain level of brain structure, a certain level of "programming" of the computer which lies between our ears, and that this program may well rule out 'Virtue is but' while letting 'Virtue is square' pass. After all, there are more ways than one in which a sentence can be a deviant one.

Shorter was writing in reply to a most instructive and also amusing article by A. N. Prior.[8] Prior's article is in part a polemic against the theory of types. He puts Russell's paradox in terms of properties, not classes, and so, instead of being concerned with the class of all classes not members of themselves, he talks of the property of nonselfinherence. But, apart from this, his solution is similar to the corollary of Quine's solution which was mentioned earlier: it can simply be false that the property of nonselfinherence is noneselfinherent (or not nonselfinherent) if there is no property of nonselfinherence. Prior wishes to say that 'Virtue is square' is not meaningless but false. That it is false follows from the assertion, which Prior would

[7] *Australasian Journal of Philosophy*, XXXIV, 1956, pp. 73-91.
[8] "Entities," *Australasian Journal of Philosophy*, XXXII, 1954, pp. 159-168.

wish to say is not meaningless but true, that virtue, being an abstraction, has no shape. Shorter, on the other hand, feels that 'Virtue is square' is not so much false as grammatically deviant and hence not a proper sentence at all. He points out that grammar has to account for what people feel when confronted with various strings of words. Some strings of words feel odd to them, as 'Virtue is but' does, and if 'Virtue is square' seems odd too, why should our rules of grammar not be strong enough to rule it out also?

The tolerant attitude to various set theories which was expressed earlier in this paper has its analogue in a tolerant attitude to both Prior's and Shorter's positions. We can envisage a formalized reconstruction of ordinary unformalized language, such that its formation rules would allow 'Virtue is square' as a correct sentence (which would be a false one), and also 'Virtue is not square' as a correct sentence (which would be a true one). We can also envisage a reconstruction which would rule out 'Virtue is square' as ungrammatical, just as it would rule out 'Virtue is but.' Of course, the latter sort of reconstruction is compatible with allowing that some of the formation rules would be deeper than others and less easily changed. In *Syntactic Structures*[9] Chomsky has argued for various levels of grammatical rules. For example, there are the levels of rules of phrase structure and of transformational rules. There is an important difference between these two levels. In order to apply the former sort of rule to a string of morphemes, one simply has to know what the string is. To apply the latter sort of rule one needs to know the steps by which the string has been derived from the initial rules of the grammar. The two sorts of rules would therefore seem to correspond to two different levels of cerebral programming. Since I am not adept at linguistics, I have in fact very little idea how to characterize the difference between 'Virtue is square' and 'Virtue is but.' Nevertheless, one must suspect that the difference is to be sought in terms of a difference in deepness of level of the grammatical rules concerned. This would connect with the fact that we can easily imagine a reconstruction of language which would allow 'Virtue is square' as a well-formed sentence, whereas it is very hard to envisage one which would allow 'Virtue is but.' Indeed, to rule out 'Virtue is square' we can rely largely on a fairly simple distinction between abstract and concrete nouns and a classification of adjectives predicable of one or the other sort of nouns or of both. Chomsky[10] characterizes such sentences as 'Sincerity admires John' (contrasted with 'John admires sincerity') and 'John frightens sincerity' (contrasted with 'Sincerity frightens John') as *semi-grammatical*, and he

---

[9] N. Chomsky, *Syntactic Structures*, The Hague, 1957.
[10] *Op. cit.*, p. 42, n. 7.

mentions the possibility of a theory of semi-grammaticalness which he has developed.[11]

It is, then, not difficult to see how to modify language (or to precisify it, since existing grammar does not seem to enable us to choose definitely between Prior's attitude and Shorter's attitude). The main obstacle to talking in the way in which Prior wants us to talk in his "Entities" lies in Russell's paradox. But Zermelo and Quine, not to mention other set theorists and indeed Prior himself in his article, tell us how to set about this. Thus in precisifying our talk about virtue we have a choice between ruling out 'Virtue is square' as meaningless and ruling it out as false. Our choice might of course have ontic implications (in Quine's sense). But this is all right. Suppose that we have to make a choice between two nonequivalent sets of axioms for a theory in physics. Then this choice has implications for what we say there is in the world, and this can be so even when there are no empirical reasons for the choice one way or the other and we make our choice on considerations of theoretical convenience and simplicity alone. In passing I might mention that this way of justifying a way of talking, namely, theoretical convenience and simplicity, disposes of one sort of objection which I once used to feel about Prior's position in "Entities." How could he have got to know such truths (if truths they are) as 'Virtue is not square'? Would this not involve a most implausible theory of intellectual intuition? This objection falls down because Prior's countenancing of such truths can be made on grounds of theoretical convenience and need involve no intellectual intuitions, at least in any tendentious sense of the word 'intuition.' So the implausibility of traditional Platonism in the light of a modern "information flow" way of looking at sense perception does not necessarily apply to a sophisticated theoretical, nonintuitionist form of Platonism. This is particularly clear in some of Quine's writings. He never defends his belief in classes by saying that he sees them with an intellectual eye. He believes in them because he holds that they are an indispensable theoretical posit for science, just as electrons are (or indeed more so).

A frequently cited case of obvious nonsense is Lewis Carroll's story of the cat which disappeared leaving only its grin (not even a grinning mouth, but just a grin). Now it is certainly very plausible to construe this talk of a catless grin as meaningless: this is the basis of Lewis Carroll's joke. But is it necessary to construe it as nonsense? Could we not alternatively construe it as a very high level truth about

---

[11] I am very well aware of my ignorance of the exciting field of modern structural linguistics. Any remarks that I have made on this subject should be accepted with caution. It is, moreover, most likely that answers to some of the tentative questions which I have ignorantly raised may exist copiously in the literature and be well known to those who are familiar with it.

grins that they do not occur on their own, but need at least a mouth and a bit of a face? If one formulated science in a Quinian way, in terms of individuals (elementary particles perhaps), classes of these, classes of classes of these, and so on (including nonhomogeneous classes), then the equivalent of the sentence 'There are no grins without cats' might even turn out to be a truth of set theory. It would, of course, contain expressions from outside set theory, but it might perhaps do so vacuously, in the way in which 'Two bananas plus two bananas equals four bananas' contains the botanical term 'banana' vacuously.

In colloquial language, as I have noted earlier, 'Nonsense!' can sometimes mean 'That's false,' rather than 'That's meaningless.' If the argument of this paper is correct, there is not a very sharp line between the spheres of application of these two meanings of 'nonsense'—indeed outside formalized languages it may not always be possible to make a clear distinction between them. (Sometimes, of course, what one man calls "nonsense" may be interpreted by another not as falsehood but as truth, as in the case of 'Virtue is not square': not having a shape, virtue *a fortiori* does not have a square shape.) It follows that there is not a sharp distinction between science as concerned with truth and falsity and philosophy as concerned with conceptual matters. Indeed, it is only on a very Baconian conception of science that anyone could want to suppose that science is concerned specifically with truth and falsehood. The beauty of science lies in its conceptual innovations rather than in its amassing of truths.

The attraction of the idea that philosophy's special sphere was the investigation of meaningfulness and meaninglessness in the various regions of discourse may partly have come, as has been noted earlier, from a preoccupation with the question of what is the proper sphere of philosophy. This preoccupation may itself have stemmed, partly at least, from the departmentalization of universities. For example, philosophers did not wish to think that they were doing lazily and amateurishly in armchairs something which was done energetically and professionally in scientific laboratories and on mathematicians' blackboards. But there is no need for the philosopher to feel like this: there is a job for the nonspecialist, even if philosophy does have the same subject matter as the sciences. There is a need for a conceptually circumspect (and hence philosophically sophisticated) attempt to see the sciences in their relations to one another, and on this basis to see what in the light of modern science is the most probable answer to some of the old questions, such as those of free will and immortality, and also to some of the new questions, such as of whether machines could think. These are questions which cannot be settled by the specialist alone because in order to answer them we need to find a synoptic view and an integrated conceptual scheme. A synthesis of the sort required depends on a knowledge of the various sciences and so it should be far removed from *a priori naturphilosophie*. Nor is it a mere encyclopedic addition of the special sciences to one another, nor is it a superscience or "governess science," to use Ryle's words,[12] which cheekily tries to tell scientists their own business. (However, there are worse vices than cheekiness, and if a philosopher does think that he could offer advice to a physicist on the merits of the Copenhagen interpretation of quantum mechanics, or to a psychologist on the merits or demerits of behaviourism, then let him offer it. It can always be rejected if it is unwelcome, with no harm done.)

In the above paragraph I have been advocating a conception of philosophy which is different from the Wittgensteinian one. The conception which I advocate can itself be accepted or rejected, and whether it is accepted or rejected is no great matter. The main thing perhaps is that philosophers should not have any cut and dried account of what their subject is. The rigid departmentalization of universities may be inappropriate to the republic of letters. As Quine has wittily remarked in conversation: "The universe is not the university." Nor need we inquire too fastidiously as to whether one or perhaps two uses of the word 'nonsense' were in question when a much revered elderly philosophical friend of mine once said: "Nonsense! I never say 'nonsense.'"

[12] *Op. cit.*

CAUSING VOLUNTARY ACTIONS

JOEL FEINBERG

> If you would worke any man, you must either know his nature and fashions and so lead him, or his ends and so persuade him, or his weakness and disadvantages and so awe him, or those that have interest in him and so govern him.
> —*Francis Bacon, "Of Negotiating"*

> It is not in our stars but in ourselves that we are underlings.
> —*Shakespeare's Cassius*

I

It is my thesis that there is no conceptual barrier, at least none imposed by common sense, to our speaking of the causes of voluntary actions. This is a view in opposition to that of Professors Hart and Honoré in their remarkable book, *Causation in the Law*.[1] These authors conclude, after a very careful scrutinizing of common sense and ordinary language, that "whatever the metaphysics of the matter may be, a [free and deliberate] human action is never regarded as itself caused."[2] I shall call this thesis of Hart and Honoré the *first cause principle*, for it asserts that every voluntary human action is a new causal start, a kind of prime mover or uncaused cause.

The first cause principle is one of three closely related but logically distinct causal principles which Hart and Honoré, but not I, find deeply embedded in common sense. For purposes of identification it should be distinguished from what I will call the *voluntary intervention principle*, which Hart and Honoré put as follows: "*the free, deliberate and informed act or omission of a human being, intended to produce the consequence which is in fact produced, negatives causal connection*."[3] Thus if an elevator operator improperly leaves an unguarded elevator shaft, and a child falls in, the operator's negligent omission can be found the cause of the child's injury; but if while the operator is absent, a

---

[1] H. L. A. Hart and A. M. Honoré, *Causation in the Law*, Oxford, 1959.
[2] The quoted passage is not from the book but from the authors' earlier article "Causation in the Law," *Law Quarterly Review*, LXXII, 1956, p. 80.
[3] *Causation in the Law*, p. 129.

third party deliberately pushes the child down the shaft, then his act takes the elevator operator off the causal hook, so to speak, severing what would otherwise be a continuing causal relation between his omission and the subsequent harm.[4] A fully voluntary act, on this view, is "a barrier and a goal *through* which we do not trace the cause of a later event and something *to* which we do trace the cause through intervening causes of other kinds."[5]

Hart and Honoré offer many persuasive examples that tend to support the voluntary intervention principle. They also concede that there are two large classes of exceptions to it. As far as I can tell they offer no more general principle explaining why the exceptions are exceptions; and since I shall be offering still other kinds of counterexamples, it will be necessary for me to explain why the supporting examples of Hart and Honoré do seem to support the principle, while the counterexamples seem to tell against it. The exceptions conceded by Hart and Honoré are first of all cases where a person persuades, induces, or entices (as opposed to merely advising or facilitating) another to do an act that causes harm, and second, those cases where one person's negligence provides the opportunity for another party's voluntary intervention. In the latter cases, the *provision of the opportunity*, for example a failure to lock up a house, can be said to be the cause of a subsequent loss, despite the fully voluntary intervention of a burglar.

The third causal principle endorsed by Hart and Honoré to which I shall be taking exception is a kind of limitation on another principle that seems to me perfectly correct. The unobjectionable principle states that causal connection between a wrongful act and a subsequent harm is "negatived" by the intervention of an abnormal or unforeseeable occurrence, a "coincidence," itself also a necessary condition of the harm. Thus, while it might be said that a wrongful obstruction of a sidewalk forcing a pedestrian into the street was the cause of his being struck by an automobile, it cannot be said that such an obstruction was the cause of his being struck by a falling airplane. That would be too great a "coincidence." So far so good; but Hart and Honoré do not give the benefit of this principle to abnormal conditions existing at the time the wrongful act commences. This limitation is expressed in what we can call the *stage-setting exclusion principle*, which Hart and Honoré put as follows: "A state of the person or thing affected existing at the time of the wrongful act (a 'circumstance') [as opposed to an 'event'], however abnormal, does not negative causal connection."[6] Common sense, they write, "regards a cause as 'intervening' in the course of events

[4] *Ibid.*, p. 130.
[5] *Ibid.*, p. 41.
[6] *Ibid.*, pp. 160–161.

at a given time, and the state of affairs then existing as the '*setting of the stage*' before the actor comes on the scene" [my italics]. The example that Hart and Honoré use in illustrating this allegedly common sense principle (which they insist is prior to and independent of legal or moral policies) is so curious that it will save me the trouble of providing my own counterexamples:

> Suppose plaintiff is run over through defendant's negligence. If on the way to the hospital he is hit by a falling tree, that is a coincidence [negativing causal connection between defendant's earlier negligence and plaintiff's subsequent death]. If, just previously to being run over, he had been hit by a tree and severely injured, that is a circumstance existing at the time of the running over and will not negative causal connection between the running over and the victim's death, even if the victim would not have died from the running down but for the previous blow from the tree. . . . It is, in a sense, as much a coincidence that the victim had been struck by a tree before he was run over as that he was struck afterwards, but the order in which the two events happen is of crucial importance in determining the consequences of the wrongful act.[7]

Hart and Honoré admit that the stage-setting exclusion principle "may be criticized as irrational,"[8] but are nevertheless resolute in imputing it to common sense. Having a rather higher regard for common sense than that, I will try to distinguish what it endorses from what it rejects in the application of the principle.

II

First I will present four stories, all of which seem to me reasonable counterexamples to the voluntary intervention principle, and the last two (at least) to the first cause principle as well. Since there are also apparently supporting examples for these principles, part of my subsequent task will be to discover the rationale of the division between the cases to which these principles apply and those to which they do not.

Counterexample 1: The Foolhardy Bank Teller. Jones, a depositor in the defendant's bank, was standing in line before the depositor's window when a bank robber entered, drew his gun, and warned "If anyone moves I'll shoot." The teller immediately grabbed something and dived to the floor. The bandit shot at him, and the ricocheting bullet struck Jones, still waiting in line, causing him severe injury. Jones then sued the bank (a more likely defendant than the impecunious bandit), charging that the teller's violation of the bandit's order

[7] *Loc. cit.*
[8] *Loc. cit.*

created an unreasonable risk of harm to the customers and that the teller's thoughtless act was thus the cause of Jones's injury. Let us suppose for the sake of this example that the teller's act was "negligent" (i.e., unreasonably risky) to the customers, and that it was a necessary condition of Jones's injury. Let us suppose also that the bandit entered the bank fully informed of the various kinds of things that might happen and fully prepared to shoot if anyone should call his bluff, and that when he did shoot, he was calm, collected, and fully aware of what he was doing. Now, on these facts, he did not intend to injure Jones since he aimed at the teller; but let us suppose that when he shot at the teller, he was aware of the danger he was creating to others, but quite indifferent as to whether anyone else was hurt or not.[9] One might well find, it seems to me, that the teller's wrongful act was *the cause* of Jones's injury even though that injury would not have occurred but for the reckless and fully voluntary intervention of the bandit.[10]

Counterexample 2: The Ingenious Suicide. Mr. Blue, tired of life, but too squeamish to kill himself, decides to use a more robust kind of person as an unwitting means. He learns of Manley Firmview, who has often announced to his friends that if he ever encountered a person who would say so and so to him, he would kill the rascal. Mr. Blue seeks out Mr. Firmview and says so and so to him. Firmview pauses for a moment, calmly considers the consequences, and then shoots the grateful Mr. Blue dead. Would common sense balk at the claim that Mr. Blue caused his own death *by means of* Firmview's free and deliberate "intervention"? That he killed himself or literally committed suicide may be going too far, but that his remarks to Firmview were the cause of his demise could scarcely be denied.

Counterexample 3: The Coldly Jealous Husband. Mrs. Green carries on an illicit love affair with Mr. Horner. Mr. Black, learning of this, and fully understanding Mr. Green's character, tells him of the facts, secretly hoping that he will kill Horner. True to form, Green investigates, learns the truth, seriously considers the matter, calmly

---

[9] Hart and Honoré hold that an act exhibiting a "reckless disregard of consequences," as well as a voluntary act intending the consequences that do result, negatives causal connection. Cf. pp. 143-144.

[10] The court in *Noll v. Marion*, 1943, 347 Pa. 213 (a case closely similar to the example in the text) held to the contrary that "while it is possible that the teller might have prevented the injury to plaintiff by remaining standing, he did nothing unlawful in attempting to save himself and his employers' property by making the choice which under the circumstances seemed best to him. . . . The cause of plaintiff's injury was not the teller's violation of the unlawful prohibition to move, but the bandit's shooting." The argument given by the court, however, is primarily directed to the conclusion that the teller was not at fault in what he did (his act was not negligent or unreasonably risky to others); whereas in the hypothetical example in the text we are to assume that the teller's conduct *was* negligent and then raise the question of causation.

decides that Horner deserves to die and that he is the one to kill him, seeks Horner out, and shoots him dead, to the great delight of the instigator, Black. On these facts, I submit, the killing of Horner by Green was a fully voluntary act, and yet common sense, for purposes of its own, might well judge that the cause of Horner's death was Black's disclosure of the facts to the cuckolded husband.

Counterexample 4: The Cocktail Party Bore. Imagine the following cocktail party conversation:

Smith: Oh, no!
Jones: What's the trouble?
Smith: George Grossfellow is here, the worst bore in the world.
Jones: What does he do?
Smith: He thinks he is a charming conversationalist, but in fact he has a stock response to almost every conversational opening, and he repeats the same self-centered stories over and over again. See for yourself. Go over and push his button. Casually mention Senator Leadwater, and I'll bet you $10 he will tell you about the time he met the great man in a railroad dining car.

Jones then does mention Senator Leadwater in a mixed group containing Grossfellow, and this causes the latter to tell his boring tale. Smith wins the bet.

One moral of these four stories is that there are many more ways of causing someone to do something than by compelling or forcing him, or persuading and inducing him to do it. One can unintentionally cause him to act by unwisely calling his bluff; or one can "get him to act" by accepting a standing offer, drawing his attention to an inciting situation, or capitalizing on his firmly fixed habits of mind.

III

Much of the confusion shrouding this subject can be removed, I think, by a distinction between two different perspectives from which causal judgments can be made—the productive and the explanatory. From the productive standpoint, a cause is always a doing or a happening. Our most primitive way of making things happen is by direct doing—raising an arm, for example.[11] Rather more sophisticated techniques involve simple manipulations of external objects. By pushing, lifting, bending, and squeezing objects we directly

---

[11] J. L. Austin thought that this was the original model of causation: " 'Causing,' I suppose, was a notion taken from a man's own experience of doing simple actions, and by primitive man every event was construed in terms of this model: every event has a cause, that is, every event is an action done by somebody—if not by a man, then by a quasi-man, a spirit." "A Plea for Excuses," *Proceedings of the Aristotelian Society*, N.S. LVII, 1956–57, p. 28.

produce changes in them.[12] But direct doing and direct producing are still pre-causal ways of making things happen, for they do not involve the use of *instruments*. One does not raise an arm, or push or pull an object by doing something still more elementary as a means; these simple motions and manipulations have a directness missing in full-fledged causal operations.

Instances of indirect producing are the clearest examples of causing things to happen. By manipulating an external object in such a way as to produce direct or primary changes in it, a person often *thereby* produces secondary changes in that or another object. We can say in this case that the manipulator causes the secondary changes *by means of* producing relatively immediate changes, or we can reserve the label " cause" for the primary changes themselves. The root notion in this use of causal language is that of a *means* or *instrumentality*. When a man causes a fire by rubbing sticks together, we can say that the friction (primary change) caused the fire (secondary change) because it was the means by which the fire was produced; and to say that friction causes fire generally comes very close to saying that one could produce an event of the latter sort by producing an event of the former kind.

Causal processes, of course, become much more complicated than simple models like stick-rubbing suggest, and as our means for producing changes become more variegated and complex, our models for understanding causal talk shift and divagate. Many simpler causal operations are mechanical and obvious: changes are initiated by pounding, pushing, or throwing objects. There is no denying that these processes have left a strong mark on our common sense notion of cause, reflected in the ready priority writers give to the impact of moving objects in their discussions of causation. "At least all of those causal chains initiated by human beings," writes Douglas Gasking, "are initiated by manipulations, that is by matter in motion ";[13] and the original "push" given even many-staged causal processes, remains persistently a model of causation in the minds of all of us.

[12] Max Black suggests that forms of direct manipulation of objects provide the most primitive casual models, although such words as "push" and "pull" he calls part of our "pre-causal vocabulary": "As we pass from the homespun language of 'making something happen' to the more sophisticated language of 'cause' and 'effect,' the influence of the paradigm remains powerful. We continue to model descriptions of cases remote from the prototypes on the simpler primitive cases, often by using metaphors literally applicable only to those clear cases. In order to understand clearly what we mean by 'cause and effect' we must labor to understand what we mean by the precausal language in which the more sophisticated vocabulary is embedded." "Making Something Happen," *Determinism and Freedom*, ed. Sidney Hook, New York, 1958, p. 20.

[13] Douglas Gasking, "Causation and Recipes," *Mind*, LXIV, 1955, p. 487. The views expressed in the text have been greatly influenced by this ingenious article.

The impact-motion model can easily be overrated, however. It may be true that causal chains initiated by persons typically begin with the moving of material bodies, but the initiating mechanism never has been exclusively that of forcing motion by striking, throwing, pounding, or pushing. Often the primary apparatus is chemical, and the initiating of motion merely the putting into position of the object to be chemically transformed—dropping it into water to be dissolved or into flame to be oxydized. In these cases there is no consciousness in the maker of exerting a force in making something happen; rather the "making" is simply the bringing together of objects in such a way that they will (surprisingly) change, the actual change-producing mechanism being independent of the maker's direct willing, and largely hidden from his eye and his understanding.

Of the various well-known techniques for causing human actions, some involve principles of even greater mystery and refinement. We can, of course, cause people to act by forms of direct bodily manipulation, or we can administer drugs or make sudden startling noises. In these cases, causing preserves some analogy to the primitive pushing or forcing model. Hypnotizing and threatening are further removed, and getting people to act by providing them with motives for acting is more special still. It seems arbitrary to deny, however, that we are causing things to be done in these cases simply because our methods bear no resemblance to the crudest productive techniques.

Hart and Honoré overstate their point, then, when they insist on " the guiding analogy with the simple manipulation of things which underlies all causal thought." That the model of causing by means of a moving force does persist to some degree, however, seems to be indicated by features of ordinary language that Hart and Honoré fasten upon. Sometimes it does seem a bit strained to say that a person caused another to act, even in circumstances where we should wish to cite his actions as the crucial casual factor in the *explanation* of the other's conduct. "Causing a person to act" easily shades into "making him act," and this suggests forcing him to act, or leaving him no choice. For that reason, we may in given cases try to avoid saying that someone or something *caused* a person to act, and say instead that *the cause of* the second person's action was so-and-so, or that the action was *due to* or a *consequence of* so-and-so. The foolhardy bankteller, for example, may deny that he caused the bank robber to shoot ("He didn't *have* to shoot," he might say) and yet admit that his own sudden motion was *the cause* of the bank robber's action. The distinction between causing and being the cause of, when recognized by ordinary language (and this is far from always the case) reflects the difference between the productive and the explanatory stand-points.

It is from the productive standpoint only that common sense

embraces the stage-setting exclusion principle. For a person to cause something to happen is always for him to intervene in the natural course of things, directly producing primary changes that result in secondary ones according to a reliable causal recipe. From this standpoint, moreover, the event that causes the effect precedes it very closely in time.

It is otherwise when we approach causal questions from the standpoint of retrospective explanation. What is to be accounted for, from this point of view, is some past event or present state of affairs, and our concern in looking for its cause is to discover why it happened or how it came to be. We *may* phrase our answer in terms of some immediately antecedent event, or of the impact of some moving body, but we *need* not. The cause, we might decide, was an event ten years earlier, or a condition, quality, disposition, action, or omission or, indeed, almost any *kind* of thing at all.

Explanations can be long stories or brief citations, and only the latter utilize the notion of "the cause." Cause-citing explanations are always occasioned by the occurrence or discovery of something out of the ordinary, what is objectively a deviation from the normal course of things or else what is subjectively contrary to someone's expectations, hopes, or fears. If the captain is almost always drunk, one might well ask what was the cause of his sobriety today, expecting a brief answer citing some single factor normally present but missing today, or normally missing but present today, but for which the surprising breach of routine would not have occurred. But if the captain is a stalwart citizen who never touches a drop, then the question "What was the cause of his being sober today?" misfires. The first mate, if asked this question by an obviously ingenuous person, might reply, "Oh, the captain is always sober; he doesn't go in for drinking." This would be to offer a short explanation, and a satisfactory one, by rebutting a mistaken presumption of the person asking the question. If the interrogator persists, however, and demands a "fuller explanation" of the captain's sobriety, no simple citation will do. The explainer will now have to tell a long story illustrating what kind of man the captain is and how he got to be that way.

When we explain something by citation, that is, by use of the formula "Its cause was such-and-such," we select out one and only one of its causally relevant conditions as more important than the others. Importance is determined by our prior assumptions, understandings, and purposes; these vary from person to person and from context to context; "the cause," therefore, is a relative thing too. For that reason, a person does not contradict himself if at one time and place, he says that "the cause" of X was Y; and in another situation he says that "the cause" was Z. Of those factors but for which X would not have occurred, Y might well be the most important for one set of purposes and Z for another; or citation of Y might contribute more to one person's understanding, citation of Z to another's.

An example might make this clear. When heavy industrial smog is trapped over a city, there is often a directly traceable effect on the death rate:

> We know very little about this effect, but we do know that air pollution does the most damage to those who have the least ability to stand any sort of stress—the old, the very young, and the sick. This was indicated in London in 1962, when, during a week-long period of heavy smog, there were four hundred more deaths than might normally have been expected.[14]

Let us imagine that Jones was one of the unlucky four hundred, that but for his severe tuberulosis he would not have died, and also that but for the heavy smog he would have recovered. Eight million other people survived the smog, a physician might reason, and but for his weakened lungs this patient should have too. Therefore, his tuberculosis was the cause of his death. An air pollution control commissioner might put it otherwise: Without the increase in air pollution this patient would have recovered; therefore, the heavy smog was the cause of his death. Of course, there is no real disagreement between the physician and the commissioner. Each has latched onto a causally necessary condition of the event to be explained which is central to his own theoretical interests and most important to his own practical pursuits. Those interests and purposes do not "contradict" one another, nor need they necessarily conflict in any way.

Often however, we are not content with this peaceful relativity. Especially when we resort to talk of "the true cause" or "the real cause," we are likely to be less tolerant of different causal citations; and when persons who agree about all the facts continue to opt for different "true causes," their causal language is a very poor disguise for a genuine opposition of purposes or policies. The cause of an ugly riot at a newly integrated southern school was the court order forcing integration, say the segregationists. No, the true cause was the inflammatory statement of the governor or perhaps the city's failure to provide adequate police protection or perhaps even the general climate of hatred, reply the northern liberals. Surely this dispute, naturally framed in causal language, is not simply over what happened, but rather over which necessary condition of what happened is the most important one for purposes of judgment, understanding, prevention, and control.

However plausible it may seem from the productive standpoint, the stage-setting exclusion principle imposes much too severe a

[14] Edith Iglauer, "Fifteen Thousand Quarts of Air," *The New Yorker*, March 7, 1964, p. 58.

restriction on causal inquiries aiming at explanation by citation. Sometimes the missing link in a person's understanding is precisely some feature of the stage setting; and sometimes one of the background conditions of an event is objectively the feature of the situation which is abnormal and without which the consequence to be explained would not have occurred. On still other occasions the causal factor of major professional concern or of focal theoretical interest is part of the setting. Moreover, the stage-setting exclusion principle would often lead to an arbitrary begging of the question when there is genuine disagreement expressed in the statement of opposed causal citations. One might inquire, for example, what caused so many men to be off the job at a factory. One answer might be that the union leaders called a strike. Another might be that the management's refusal to budge from a negative bargaining position was the true cause. This is a familiar and difficult kind of disagreement. If we approach it strictly from the productive standpoint, however, and apply the stage-setting exclusion principle, its resolution is easy. The impasse in negotiations was stage setting; the union leaders' order to strike was an active intervention on this stage and therefore the cause of the strike. Of course, from one point of view, there is nothing objectionable in saying that the union leaders, by giving their order, caused the strike; that is how strikes are "made to happen." But if we decide therefore that the unionist is forbidden by "common sense" from citing the management's failure to accede to union demands as "the real cause," we are taking away his tongue. He will now have to say that the management's refusal to bargain was not the cause of the strike but only the most important of all those conditions in the absence of which the strike would not have occurred! Similarly, the Negro children's "intervention" on the stage so carefully set by the segregationists would have to be called the cause of the riot, and the rioters win their point, as it were, by default.[15]

---

[15] Compare the arguments put forth by Bertrand Russell and others to show that President Kennedy and not Premier Khrushchev caused the Cuba crisis of 1962. When Kennedy declared a naval blockade at the time Soviet ships were on their way to Cuba, according to this line of reasoning, he was intervening on a stage part of whose setting was the presence of Soviet rockets in Cuba. His intervention immediately led to the crisis. On this view, I suppose, if war had ensued, it in turn would have been caused by Khrushchev for the American blockade then would have receded into the causal background against which a Soviet firing on the blockading ships would be the active intervening cause. Thus the stage-setting exclusion principle, in cases like this, turns into a version of the doctrine of the "last clear chance," often applied in law courts. Common sense, I would submit, would not necessarily select the last active intervention, in seeking the cause of the crisis, but rather that act or event that was a radical deviation from routine, and that was clearly the Soviet construction of missile bases in Cuba. But see Bertrand Russell, *Unarmed Victory*, New York, 1963, especially pp. 27–29 and 32–65.

If we remember that the point of an explanation by causal citation is to induce understanding, we will be less inclined also to accept the first cause and voluntary intervention principles. We may smile when told that the cause of a grown man's voluntary action was some feature of his infant toilet training, but such a judgment, if based on a correct assessment of the facts, would commit no conceptual solecism. If it is true that but for the manner of his toilet training he would not have acted as he did, we can imagine interests and purposes for which that fact has quite sufficient importance to qualify as "the cause" of his subsequent voluntary behavior. Yet it would be misleading in the extreme (if there were anyone so naïve as to be misled) to suggest that his mother thirty years earlier "made" him act as he did today. "The cause" in this example is not the *instrument* by which he was "made" to do what he did; rather, it is the *light* by which we come to see and understand his action more clearly.

Similarly, if our goal is understanding, we will not hestitate to trace a puzzling event right back through one, two, or many fully voluntary acts done with the intention of bringing it about, to a much earlier factor, more obscure perhaps, but equally necessary, and much more interesting. So some have said that the cause of World War II was the unfair Versailles Treaty; and others have found the cause of the Protestant Reformation in Julius Caesar's failure to conquer the German tribes.

The examples of Hart and Honoré that seem to show that common sense endorses the voluntary intervention principle owe their persuasiveness to the operation of a quite different principle, one endemic in the explanatory standpoint, namely, that a highly abnormal or otherwise especially interesting occurrence, whether a human action or not, negatives causal connection between an earlier act or event and a later upshot. It is certainly correct, for example, that "a diner's death is not said to be caused by, or even a consequence of, the laying of the table from which a murderer seized the fatal knife";[16] but this is not because the murderer's act is a "voluntary intervention." Rather it is because it is the abnormal deviation that distinguishes the whole incident from other dinners where diners are *not* killed. If the murder occurred in a prison dining hall, or a mental hospital, where knives are never set on tables and diners may be expected to get violent, then the laying of the table would be the abnormal event of great explanatory power, and the provision of opportunity "the cause." The pertinent principle here is that *the more expectable is human behavior, whether voluntary or not, the less likely it is to "negative causal connection"*; and when the stakes are high,

---

[16] The words quoted are from Philippa Foot's perceptive review of *Causation in the Law, Philosophical Review*, LXXII, October, 1963, p. 510. The example is found in Hart and Honoré on p. 66.

as in our bank robbery example, consequences will be traced right back through a voluntary act, providing only in retrospect it seems "not highly extraordinary"[17] that it intervened.[18]

IV

Among the more familiar examples, from the productive point of view, of causing things to happen, are those of *triggering, precipitating,* or *igniting.* The causal instruments employed by these techniques are rather more sophisticated than the elementary tools of the simpler causal models, consisting either of machines (locks, triggers, catches, and springs) or else chemical operations (heating, cooling, mixing reagents). The following elements are generally characteristic of these complex causal processes: (1) In each case there is some point (a threshold) at which a gradual accumulation of small and unnoticed quantitative changes suddenly yields a large and conspicuous transformation. Pressure on a trigger builds up until it unlocks or releases a spring; the temperature of a liquid is gradually lowered until it suddenly yields a solid precipitate; the temperature of an explosive is gradually raised until it ignites. (2) The person who initiates the causal process may have to do a great deal of work first to build the mechanism up to the threshold point, or he may find the mechanism ready and cocked, needing only the slightest "final push" to go off. (3) The causal mechanism may itself be either a simple machine or an interlocking assemblage of mechanical or chemical parts many of which themselves exploit threshold phenomena. In complex machines there is something like a chain reaction of triggered responses between the operator's initiating movement and its final upshot. (4) There is a great gain in efficiency. The operator does very little work in pushing a button or lighting a fuse; the intervening machinery yields a "stepped-up output," or response out of all proportion to the energy originally expended. (5) Finally, it is characteristic of these systems that once the process is initiated, through at least one organic phase it is irreversible.

[17] This is the phrase used in the *Restatement of the Law of Torts* (Sec. 447). A negligent defendant will be liable for harm, according to the *Restatement*, despite the intervening act of a third person if "a reasonable man knowing the situation existing when the act of the third person was done would not regard it as highly extraordinary that the person had so acted."

[18] There may be still another explanation for the persuasiveness of some of the examples by which Hart and Honoré support the voluntary intervention principle. In many of them an intentionally wrongful act intervenes between a negligently wrongful act and the harm for which both were necessary conditions. Intentional wrongdoing is usually regarded by moral common sense as more serious than negligence, and it is very difficult to keep separate the questions of moral blameworthiness and causation. This is a point that has not escaped the attention of these authors, however. See their vigorous counterargument, pp. 77f.

Once the rifleman has released the hammer by sufficient pressure on the trigger there is no preventing a bullet from emerging from the barrel and travelling hundreds of feet. The whole physico-chemical stepping-up process is faster by far than the eye or hand of whoever put it in motion.

Now when the initiator has very little to do to put the process in motion—when by merely pushing a button he provides the final straw or the last push—then from the *explanatory standpoint* his action may be altogether uninteresting. Of course, if we know all about the machine, yet wish to know in this instance how an end product came into existence, we may readily settle for a simple citation explanation: The cause was Jones's pushing of the button. But we may *see* Jones push the button and still not understand how the end product came about. After all, mere finger pushings are familiar parts of the everyday scene, but rarely are they followed by such extraordinary creative outbursts. In this case we may not settle for any citation explanation. We may need to hear a very long story indeed to finally acquire understanding. Or, if we are mechanically sophisticated, we may settle for a citation explanation which mentions some one special feature of *this* machine. The last thing we would settle for, in any case, would be the trivial commonplace of the button pushing.

Yet, from the productive standpoint, the operator's pushing of the button is the purest prototype of causing something to happen. The machinery is all stage setting or background against which the newly intervening force of the moving finger sets things going. In short, what clearly causes, or makes something happen, from the productive standpoint, may not be worth mentioning when we look for "the cause" from the standpoint of retrospective explanation.

We often use the triggering, precipitating, and igniting metaphors in accounting for human actions, even, I would argue, for fully voluntary ones. A simple remark, a mere gesture, a routine action of all apparent innocence, may bring about an enormously stepped-up response. Triggering metaphors seem especially appropriate in describing these processes, partly because, as P. B. Rice once put it, "All living beings are loaded and cocked when they come into existence"[19]—i.e., with governing biological propensities; and partly because experience gradually makes its mark on us, building up by little increments to the threshold of big response. Propositions gradually grow more credible to us until suddenly, in a quite ordinary setting we reach the threshold of belief, as in a sudden flash. Similarly, expensive objects become gradually more tempting until suddenly a disproportionately trivial stimulus triggers a wholesale depletion of our bank account. So also a man struggling

[19] Philip Blair Rice, *The Knowledge of Good and Evil*, New York, 1955, p. 178.

with his temper may have it "ignited" by a very small incident or remark.

It happens commonly enough that one person, either knowingly or unintentionally, triggers the action of another. The bank robber was all wound up to shoot, so to speak, at the slightest provocation; the teller perhaps inadvertently but effectively provided that slight stimulus. Now when this sort of thing happens we may or may not cite the triggering action as "the cause" when we give a retrospective explanation. If it is well known that the second actor had the disposition to act as he did in the circumstances, or that his condition was very near the relevant threshold, then the request for an explanation may ask, given that condition, what made the difference this time, and a satisfactory answer will cite the "triggering" cause. But the triggering stimulus may seem the "normal" element in the situation to one unacquainted with the dispositions and threshold conditions of the person responding, in which case "the cause" will not be the stimulus but rather some aspect of the "trigger mechanism." This is another illustration of the relativity essentially residing in the explanatory perspective. But from the productive standpoint the causal issue is much more clear-cut and objective. Triggering another person's action is a very familiar way of "causing things to happen." Mentioning something to a person, for example, may remind him of something that otherwise should have remained just below the threshold of his attention, and given his governing habits, he may promptly speak up and recite what he has been reminded of. This is the way the sport in our earlier counterexample caused the party bore to tell his oft-told tale.

Now the crucial question arises: If A's action is triggered by B's and especially if it is caused to happen by B's intentional triggering, how can it possibly be said without qualification to be A's own action, a free, informed, deliberate, i.e., "fully voluntary" act? To begin with, I should like to concede that not all acts triggered by other persons can be said to be voluntary. Some triggers are extremely hard to pull and when the required pressure is great enough one may have to strain mightily to set things going. When a person's threshold is high and his relevant condition low, a would-be instigator may have to build it up gradually over an extended period of time by coaxing, playing on emotions, "fanning," "putting on the squeeze" before the desired act can be triggered. When this is the case we would be reluctant, of course, to call the triggered act fully voluntary;[20] all the more so when the triggered disposition itself

---

[20] There are perhaps some exceptions to this. Often triggering is extended *stimulation*. When the stimulated response is not a happy one, we call the stimulation *provocation;* when the response is noble, the stimulation is called *inspiration*. Sometimes it takes a great deal of inspiration to reach the threshold of heroic action or

was implanted by the instigator through indoctrination, hypnotic suggestion, or other direct and prolonged manipulative techniques. Further, when the dispositions in question, however acquired, are regarded as diseased or unhealthy, the actions they yield when triggered will not, without severe qualification, be held to be voluntary.

This, however, still leaves us a large range of easily triggered, naturally acquired, and psychologically normal dispositions and threshold points. I should like to suggest that the more important to a full and comprehensive understanding of a triggered action are the actor's own dispositions and thresholds, the more likely we are to consider the act truly his, providing those dispositions and thresholds are not biologically or psychologically abnormal, and provided that they were not imposed on him by manipulation. When the trigger-pull is easy, and only the slightest commonplace sort of stimulus is required to set off an action, then the major part of the explanation of that action will necessarily involve the agent's own complicated self; and the more the triggering diminishes in explanatory significance, the less reluctant we are to regard the act as voluntary.

These points can be illustrated by a story a colleague used to tell to reveal what he took to be the incompatibility of determinism with common sense. Imagine that there are fine invisible wires connected to your body in various ways and attached to an elaborate machine somewhere so that whenever the machine operator pushed a certain button, your left arm would rise, and whenever he pushed another, your right arm would rise, and so on. Now, if you did not know of this arrangement, my friend would argue, you might think that you were raising your arms and doing all the other things you do voluntarily. You may be aware of certain purposes, motives, and apparent "volitions" which foster this illusion, but once you learned the truth, you would abandon the belief forever as an ignorant conceit. At this point someone would always protest that determinism does not imply that there is some *person* deciding at every moment how my body is to move. Well, change the example, then, my friend would reply. Imagine that the invisible wires are hooked up to such things as weather vanes, so that, for example, when the wind comes from the north you would move your left arm, and when it is southerly, your right, and so on. It does not matter *how* we are plugged in to external nature, my friend concluded; insofar as we

---

difficult achievement; but in these cases we do not characterize the resultant act as less than voluntary. The point in the text does, however, seem to apply to all "extensively stimulated" bad actions. In this connection it should be noted that triggering is literally *releasing* a lock or catch; and the "lock" or "defenses" of a human being are, after all, a part of him. When it takes a great deal of stimulating for a person to respond, the stimulation is less a releasing than a kind of *overcoming*.

are plugged in, our actions are to that extent not our own, but rather the doings of that, whatever it may be, into which we are plugged.

The reply to this argument, it seems to me, is that it does make a good deal of difference not into *what* we are plugged but rather *how* we are plugged in. If the determining influences are filtered through our own network of predispositions, expectations, purposes, and values; if our own threshold requirements are carefully observed; if there is no jarring and abrupt change in the course of our natural bent; then it seems to me to do no violence to common sense for us to claim the act as our own, even though its causal initiation be located in the external world. In short, the more like an easy triggering of a natural disposition is an external cause, the less difficulty there is in treating its effect as a voluntary action.

On the other hand, if I were plugged into nature in such a way that the determining influence bypassed my own internal constitution, failing to utilize its latent tendencies, then clearly the resultant act would not be voluntary. If I were a guest at a formal dinner party, and some cosmic button pusher caused me to pass up my favorite dessert and, contrary to all my tastes, inclinations, and scruples, chew hungrily on the discarded cigarette butts in a nearby ashtray, I would surely be taken as the victim of some insane impulse or violent seizure. Triggering another person's actions need be nothing at all like this.

V

Before concluding, I should like to consider briefly a provision of our criminal law for which the causation of voluntary behavior raises serious problems both conceptual and moral, namely, that which allows an accused person, in some circumstances, the defense of "entrapment." When this defense is available to a defendant, he will be acquitted if he can prove that his alleged criminal conduct occurred in response to the "inducement or encouragement" of a police official or his agent.

No one would think of depriving the police of all initiative in trapping criminals. The plainclothes policewoman who mingles with a department store crowd in order to lure purse snatchers, for example, does not tempt honest people to depart from the path of rectitude. She lures the already determined or habitual criminal. Her behavior, perfectly normal and unobtrusive in every other way, lures precisely one class of persons to whom she provides an opportunity to do what they are already bent on doing. There are some crimes that are almost impossible to detect without such techniques—"bribery, prostitution, and the illegal sale of narcotics, liquor, and firearms seem the most prominent."[21]

[21] "Note on Entrapment," *Harvard Law Review*, LXXIII, 1960, p. 1338.

At the other extreme are clear examples of police abuses. If a policeman incites a person to crime either by deceiving him into believing the criminal act is legal or by threatening harm, then the actor's criminal act is involuntary, and without question, he may plead the entrapment defense. But between the extremes of passive allurement on the one hand, and coercion and deception on the other, are a range of techniques that cause many problems. Police agents in plain clothes have caused persons to act criminally by suggesting, inviting, advising, requesting, urging, coaxing, imploring, enticing, and so on. It is difficult to know how to draw the line between the proper and improper employment of such methods.[22]

But the most difficult kind of problem posed by entrapment techniques involves the following kind of case. Suppose that there is such a high degree of "readiness" in a certain person to commit a certain criminal act, that only the slightest degree of inducement is required to bring him to the threshold. Perhaps a mere insinuation, an intriguing remark, or an appealing and persuasive tone will do the trick. Since these techniques, unlike insistent requesting, exhorting, seducing, and the like, are commonplace stimuli of generally small effectiveness, they would almost certainly preclude the entrapment defense. But suppose that, but for the policeman's use of these techniques, it is highly unlikely that the person incited would ever have committed the criminal act. He may have remained for the rest of his days "ready" to commit the crime, that is, in a condition such that it would not take much to induce him to commit it, and yet never cross that threshold. This is an example of causing a generally law-abiding citizen to commit a crime he would not otherwise have committed and yet depriving him of his entrapment defense.

---

[22] The Model Penal Code rule (American Law Institute's final draft, 1962, Sec. 2.13) forbids techniques likely to induce to the performance of a criminal act "persons other than those who *are ready* to commit it" [my italics]. This will hardly yield a workable test of entrapment, however, if the test, in turn, of a person's readiness is whether it would take a large or small amount of inducement to get him to act. Note the unavoidable analogy to triggering. It is as if we were to say that the test of whether or not a trigger mechanism was in a state of prior readiness is whether or not it took an inordinate amount of pressure to release it; and the test of whether the triggering pressure was "inordinate" is whether it would have been sufficient to trigger even "unready" mechanisms.

The Model Penal Code criterion, then, like most legal tests, has only a specious precision. Interpreted as a kind of litmus test, it is imperiled by circularity. It had better be interpreted as a rough guide to the kinds of consideration relevant to the question and the way these considerations (acts of policemen, predispositions of the inducee, predispositions of other people) are interrelated. The test says, in effect, that the defense of entrapment is not available to those whose prior condition was already *very near* the threshold of criminal action, with "very near" left necessarily vague.

Since in cases of this kind, the police inducement, however small, is a necessary condition of the criminal act, it seems fair to say that the criminal so induced will be punished for his *predisposition*, that is his readiness, to commit the crime, or put another way, for his high susceptibility to inducement. The intuition that such punishment is not fair is disconcertingly difficult to support with argument. Perhaps much of it is derived from the notion that a legal system promises to punish only acts, and in cases like this the spirit, if not the letter, of that promise is violated. But to circumvent that difficulty let us discuss the justice of a hypothetical general practice. Imagine a system of law enforcement in which police agents search out people who teeter on the brink of criminality, and who may or may not eventually fall in, and deftly provide them with the final ever-so-slight stimulus to criminal action. Would this be a just or unjust general practice? We are, of course, understandably loathe to confer powers on policemen that can be misused, and this might be a quite conclusive reason against the scheme under discussion. But would the scheme be unjust to those persons sought out and trapped? Insofar as they are punished only for their predispositions, are they being treated unfairly?

In support of a system of anticipatory enforcement one might cite its effectiveness as social prophylaxis. After all, it is socially dangerous to have loaded and cocked weapons lying around, especially those with delicate trigger mechanisms. Why not seek them out and detonate them harmlessly? And if this involves neither coercion nor deception of free and responsible moral agents, how can it be unjust to them?

It might be said that mere predisposition does not constitute sufficient *culpability* for just punishment, but from the moral point of view, there appears to be no significant difference between the person of high readiness to crime who is induced to act and the person of high readiness who never finds his inducer. The locus of moral culpability in each case *is* the predisposition, and what distinguishes the two cases is mere luck; for it is not to the credit of the one that he did no criminal act if it was only an accident that he was never brought across the criminal threshold. Furthermore, the ready criminal is no less guilty for being induced to act by a policeman in disguise than thousands of others of like disposition who are induced by private persons.

This problem about justice is far too difficult to pursue any further here. The thesis I wish rather to defend is a very modest one: that the problem could not even be formulated without assuming that caused behavior can be fully voluntary. More specifically, the possibility of caused voluntary actions is what gives the moral problem of anticipatory enforcement its difficult and dilemmatic character. Insofar as another party gets a person to do something he might otherwise never have done, we find it unjust to hold that person responsible. The initiative, we say, was with the other party. On the other hand, insofar as the induced act was freely done by a person who was in a high state of readiness to do it, we find that it was truly his own action, and one for which he is fully responsible. There is no conceptual difficulty, I submit, in this description of an act as caused and voluntary; but that is exactly what, from the moral point of view, causes all the trouble.

COMMENTS

KEITH S. DONNELLAN

What Mr. Feinberg calls the *first cause principle* states, in the words of Hart and Honoré, that "whatever the metaphysics of the matter may be [a free and deliberate] human action is never regarded as itself caused" (p. 29). Mr. Feinberg attacks the principle by citing counterexamples, situations in which we should both hold that a certain action is free and deliberate and at the same time speak of what caused it. I believe that he has shown that such counterexamples can readily be found. If the dispute over this principle is ultimately over whether or not we should talk of what caused a person to act, although the act be fully voluntary, then I believe Mr. Feinberg wins. But I think the dispute is deeper than this. Mr. Feinberg, after all, goes on later in his paper to discuss voluntary actions in terms of "triggering," "thresholds," and "dispositions to act." He moreover relates this discussion at one point to the problem of free will and determinism. It does not seem to me that winning the earlier battle fully entitles him to this as the fruit of his victory.

I believe, then, that what Mr. Feinberg can claim as against Hart and Honoré is that we not only use "many important causal idioms" such as 'because of' and 'as a result of' in connection with voluntary actions, as they concede,[1] but the word 'cause' itself is used. The bank robber in one of Mr. Feinberg's cases may say that the teller's movement caused him to fire, and the cuckolded husband may say that Mr. Black's disclosure of the facts caused him to seek out Horner and kill him. It is, by the way, not only the deeds of other people which are said to cause voluntary actions: the ringing of the chimes may cause me to hurry to class and the sight of rain clouds may cause me to seek shelter.

But having conceded this much, it does not seem to me that we must also admit that there are no important differences between the causes of voluntary action and the causes of other events. I believe that behind the somewhat incautious statement which Mr. Feinberg quotes from an earlier paper of Hart and Honoré is their insistence that "In this field of relationship between two human actions we have to deal with the concept of *reasons* for action rather than the causes of events."[2] They argue for the, by now, familiar contrast between explanations in terms of causes and explanations in terms of

---

[1] H. L. A. Hart and A. M. Honoré, *Causation in the Law*, Oxford, 1959, p. 49.
[2] *Ibid.*, p. 48.

a person's reasons for acting. It is possible that their insistence upon this distinction has prevented them from appreciating the possibility that we may use causal language in connection with a person's reasons for acting. But it seems to me that they can agree to this and without prejudice to the distinction they want.

As a first shot at an analysis, we might say that when we speak of the cause of a voluntary action we mention something the awareness of which provided the agent with his reason for acting as he did. It seems very likely that this is significantly different from citing the cause of a "natural" event and, possibly, the difference will make Mr. Feinberg's model of triggering inappropriate.

If, for example, the triggering model is taken seriously, then having cited the cause of a voluntary action it would seem that we might be interested in how that triggered the action, interested, that is, in the causal mechanism. Having learned that pouring insecticide in the river caused the fish to die, we may ask *why* that caused them to die? (The answer might be in terms of the effect of the insecticide on their gills.) Having learned the cause of a voluntary action we do sometimes ask why it caused the person to do what he did. "Why did the teller's movements cause the bank robber to fire?" "He had warned them not to move and he thought the teller was reaching for an alarm button." But the answer here appears to give us the reason which the awareness of the cause provided the agent, rather than intermediate causal stages through which the cause produced its effect. (The difference is perhaps mirrored in what seems to me to be a fact about ordinary usage, that in the case of natural events, the "why?" question can be paraphrased by a "how?" question, e.g., "*How* did that cause them to die?" But it would seem odd to do this when it is a voluntary action, e.g., to ask "*How* did the teller's movement cause him to fire?")

Hart and Honoré list several other differences which they find between what we may now call *causes* of voluntary actions and *causes* of other events. Among these is, for example, that in citing something as the cause of a voluntary action we are not committed to any sort of generalization to the effect that this usually causes people to act in this way or that this person usually acts in this way when such an event occurs. But I should like to mention a difference which they do not list, but which seems to me both to have a bearing on their ultimate concern, responsibility, and on Mr. Feinberg's "triggering" model.

I think we must admit, and this appears to be support for Mr. Feinberg's position, that if what one person has done is the cause of another's voluntary action, that person may be liable for censure or praise if evil or good is the result. Thus, certainly Mr. Black would be morally culpable for disclosing the facts to the cuckolded husband, knowing that the latter would very likely do harm to the lover. Even the bank teller may be blamed for acting foolishly in the face of a threat by a bank robber and held censurable by the victim of the ricocheted bullet. Of course the principal agents in these tales are not absolved thereby. But when we say that Mr. Black or the bank teller may be held accountable, we do not, I think, mean that they can be held accountable by the cuckolded husband or the bank robber. Neither can *accuse* those who provided them with reasons for acting of causing them to act wrongly. It would be absurd for Mr. Green to blame Mr. Black for causing him to seek out Horner and kill him. On the other hand, if either action were to be judged *not* voluntary, then I think grounds for complaint might be in order against those who provided the cause.

Now there are cases in which one can blame another for doing something which caused his voluntary action. If I provide you with false information and you act upon it, subsequently regretting your action, you may blame me. Also, if by my action I produce a situation which gives you a choice of evils, you may blame me. But the principle I wish to state may be put so as to take care of these possibilities. If X's deed is the cause of Y's voluntary action, Y may still come to regret that, even in the light of what X did, he did not act in some other way or refrain from action altogether. So the bank robber may come to repent his act of firing. And in blaming himself he does so despite the truth of the statement that the teller's movement caused him to act. So, too, a man may think his action unwise even granting that he had false information or was put into a situation of choosing between evils. And it is here that he cannot level blame against the person who provided him with his reason for acting. Of course all of this is simply a corollary to the principle that if the act was voluntary then no matter what its cause, there were alternative courses of action. But putting it in terms of culpability helps to show why Mr. Feinberg's triggering model needs a great deal more support than I think he has given it if he means it to be a model for our common sense beliefs. On his model it seems to me inexplicable why one should not blame another for triggering his mechanism or taking him over the threshold even though the action is voluntary. Nor will it do to say that what I blame myself for in these cases, and not you, is having such a low threshold. A case in which this analysis is appropriate would be this. Suppose that I am trying desperately to give up smoking and have a very low threshold of resistance. Now if you, knowing all this, put out cigarettes, offer me a smoke, etc., and I succumb, I may very well blame you and, I think, exonerate myself from blame for this lapse. But I may, of course, blame myself for having ever gotten into such a state. Here I blame you and not myself for the particular action even though I can blame myself for having such a low threshold. But insofar as I do blame you and not myself I think I am committed to the action's not being wholly voluntary.

What I have done is to suggest that Mr. Feinberg has not ruled out the possibility that even though we do speak of the causes of voluntary action, there are yet significant differences in our treatment of these causes and those of natural events. I am not suggesting that an appeal to what we say or do in the two cases would show that no triggering model is appropriate in the case of a voluntary action. It is rather Mr. Feinberg who suggests that what we say and do *supports* such a model and it is then appropriate to ask whether this is so.

COMMENTS

KEITH LEHRER

My comments will consist in the defense of two theses. In the first place, I shall argue that Mr. Feinberg has misinterpreted the voluntary intervention principle and consequently that his counterexamples miss their mark. Secondly, I shall argue that even if it is consistent to say that a voluntary human action is caused, there is still reason to doubt that determinism is consistent with our common sense beliefs about voluntary human actions.

Let us consider the *voluntary intervention principle*, namely, that "the free, deliberate, and informed act or omission of a human being, intended to produce the consequence which is in fact produced, negatives causal connection" (p. 29). To understand this principle, we need to bear in mind a distinction drawn by Hart and Honoré between two kinds of causal statements—explanatory causal statements and attributive causal statements. Explanatory causal statements explain why something puzzling has happened. Attributive causal statements attribute some harm or benefit to the action of some person who is said to have caused it. Sometimes the action to which something is attributed will not explain what puzzles us, and sometimes the action which explains what puzzles us is not the action to which some resultant harm is to be attributed. When Hart and Honoré say that a voluntary human action negatives causal connection, they must be understood as asserting *only* that the intervention of a voluntary action nullifies the *attribution* of the intended consequences of that action to any earlier action. They are not asserting that no earlier action can be part of the causal explanation of the later consequences.[1] Contrary to what Mr. Feinberg suggests when he says, "If we remember that the point of an explanation by causal citation is to induce understanding, we will be less inclined also to accept the first cause and voluntary intervention principles," (p. 39) the voluntary intervention principle has nothing to do with causal explanation. What is negatived according to this principle is not causal explanation but only causal attribution.

This point can be further clarified by considering some of the four restoration stories Mr. Feinberg tells, the alleged counterexamples to the principle in question. In the case of the coldly jealous husband, it is quite clear that the death of Horner is to be attributed to the

[1] H. L. A. Hart and A. M. Honoré, *Causation in the Law*, Oxford, 1959, pp. 22-23 and pp. 59-78.

husband, Mr. Green, and to no one else. It surely would be a mistake to attribute his death to Mr. Black. On the other hand, some action that occurred prior to Green's action, for example, Black's disclosure of the facts of infidelity to Green, may serve to remove our puzzlement over the fact that Green has murdered Horner and, thereby, causally explain the latter's death. The voluntary intervention principle does not deny that Black's action might serve as a causal explanation of Horner's death; it only denies that Horner's death may be attributed to Black's action. Similar remarks could be made to show that the other stories are not obvious counterexamples to the principle in question.

I conclude, then, that Mr. Feinberg has failed to produce a counterexample to the voluntary intervention principle because he ignores the distinction between explanatory causal statements and attributive causal statements, a distinction that is essential to a proper understanding of the principle Hart and Honoré wish to defend. Of course, it may well be that the distinction between explanatory causal statements and attributive causal statements is obscure or otherwise untenable, but Mr. Feinberg has not proven or even argued that this is the case.

However, I believe that Mr. Feinberg is not only interested in proving Hart and Honoré mistaken, he is also interested in showing that there is no incompatibility of determinism with common sense. That is, he wishes to prove that our common sense beliefs concerning human action are consistent with determinism. I now wish to argue that even if there is no inconsistency in saying that a voluntary human action is caused, it does not follow from this that determinism is consistent with our common sense beliefs concerning voluntary human actions. To illustrate this point, I shall read you a short passage from John Locke. In his rather remarkable English prose, Locke once wrote,

> Suppose a man be carried whilst fast asleep into a room where is a person he longs to see and speak with, and be locked there fast in, beyond his power to get out; he awakes, and is glad to find himself in so desirable company, which he willingly stays in, i.e., prefers his stay to going away. I ask, is not this stay voluntary? I think no one will doubt it: and yet, being locked fast in, it is evident he is not at liberty not to stay, he has not the freedom to be gone.[2]

I shall refer to the man in this story as "Mr. Lockedin." The moral of his story is this: Mr. Lockedin's stay in the room is voluntary even though he could not have done otherwise.

To see that Mr. Lockedin's stay in the room is voluntary, it is important to notice that he does not know that he is locked in the

---

[2] John Locke, *An Essay Concerning Human Understanding*, New York, 1959, Vol. II, p. 317.

room; that is why Locke says that he is carried there while fast asleep. To make the example as convincing as possible, we may suppose that the man believes that he can leave the room and that he considers leaving the room but chooses to stay. With these suppositions, it is quite clear that when the man stays in the room, his stay is voluntary. But we have not yet won the day for the compatibility of determinism and common sense. For it is also correct to say that, the door being locked, Mr. Lockedin could not have done otherwise. This shows that the statement that a human action is voluntary does not entail that the agent could have done otherwise. Therefore, even if the statement that a human action is voluntary is compatible with determinism, it does not follow from this that the statement that the agent could have done otherwise is compatible with determinism. And that is precisely the rub, because it is a common sense belief, not only that some human actions are voluntary, but also that sometimes the agent could have done otherwise. The latter cannot be shown to be compatible with determinism by showing that the former is.

Moreover, there seem to be very compelling reasons for thinking that the statement that an agent could have done otherwise is not compatible with determinism. In the first place, even if the statement that a person could have done otherwise is consistent with the statement that his action was caused, it would not follow from this that the former statement is consistent with determinism. The reason for this is that determinism implies not only that every human action is causally determined but also that the conditions that determine each action are causally determined, that the conditions that determine those conditions are causally determined, etc. On the usual interpretation of determinism, this chain of causal determination extends backward indefinitely in time, and the relation of causal determination is transitive. Therefore, it is a consequence of determinism that every human action is causally determined by conditions that existed before the agent was born and, consequently, over which he had no control.

Thus, to prove the consistency of the statement that a person could have done otherwise with determinism, we must prove that the former is consistent with what is entailed by determinism, namely, that the agent's action was causally determined by conditions which existed before he was born and over which he had no control. To prove the latter would be a formidable task indeed.

REJOINDERS

JOEL FEINBERG

One important modification of my view is necessary. Throughout my paper I speak of "the cause" as an interesting or important *necessary* condition simply. I should be better advised to mean by "the cause" of an event generally, "a condition which, when conjoined with circumstances normally present, is *sufficient* to bring about events of the type in question," and to mean by the cause of a given event, "a condition which, when added to circumstances already present, is *sufficient* for its occurrence." This emendation detrivializes my thesis without diminishing its plausibility, for there is ample support in common sense for the view that some voluntary actions are caused even in this strong sense and, in particular, that they are often caused by the actions, intentional or unintentional, of other people. A much less dramatic example than those in my article might help make this clear. I can get an acquaintance to say "good morning" by putting myself directly in his line of vision, smiling, and saying "good morning" to him. My doing these things is not only a circumstance but for which his voluntary action would not have occurred, it is also a circumstance which, when added to those already present, "made the difference" between his speaking and remaining silent. So much, at least, is conveyed by the phrase "getting him to act," which would not be used if the inciting behavior were not considered a *sufficient* cause.

Mr. Donnellan apparently accepts my main thesis but warns against claiming the truth of certain controversial philosophical doctrines as "the fruit of [my] victory." His warning is well heeded. I do not wish to claim that man is but a machine or that human dispositions to act are literally "trigger mechanisms" or that "there are no important differences between the causes of voluntary action and the causes of other events." I strongly regret that my lack of clarity in stating my intentions conveyed such impressions. All I meant to do in the section Mr. Donnellan criticizes was to cite the commonplace that we do often use triggering metaphors in describing some voluntary human actions and to argue for the appropriateness of the metaphors (as metaphors), which still seems to me beyond question. I emphasized certain similarities between the human actions for which these metaphors are appropriate and mechanical contraptions like guns, but I did not intend to suggest that there are "no important differences." F. Scott Fitzgerald is alleged to have insisted once to Ernest Hemingway that there are important differences between the rich and other people, to which Hemingway is supposed to have replied, "Yes, the rich people have more money." Human beings are importantly different from mechanical contraptions in that (among other ways) they have intentions, purposes, values, reasons, and goals; and I have no more desire to underrate the significance of these things than did Hemingway the importance of money.

Because human beings are different in these ways from mere machines, we can ask questions about their behavior of a sort that do not apply to events of other kinds. We can ask, for example, "*Why* did he do that?" meaning, "What were his reasons for doing it?" That this is a question appropriate to actions but not to other events, I readily grant; but it would be incorrect, I believe, to infer from this the converse, that questions of a form readily applicable to mere events are not also applicable to voluntary actions. We can ask, I submit, of anything that has come to be, *how it came to pass* that it came to be. Sometimes an adequate answer to a "How did it come about that . . . ?" question is a simple statement of a person's reasons for acting; but sometimes such an explanation fails to resolve perplexity, and one must cite the agent's remoter goals or purposes or his motives or his habits, policies, or "springs of action" or even a general account of "the sort of person he is." Moreover, while an explanation in terms of reasons for acting does render a voluntary action more intelligible (at least in one dimension), it might be maintained that some actions, at least, do not become wholly comprehensible until one knows, for example, *how it came about that* these reasons "weighed with him"[1] or that he had these purposes and not purposes of another kind or that he was this sort of person and not some other type. Satisfactory answers to the latter questions, of course, will not mention further reasons or purposes of the agent, but will instead possess a form that assimilates persons and their actions to the rest of nature. In short, while I must concede to Donnellan that "What were his reasons" questions cannot be *paraphrased* as "How did it come about that" questions, still answers to the latter might supplement or supersede answers to the former, and explanations in terms of reasons might well be subject to a kind of integration with explanations of other kinds.

In one place, Donnellan puts his point in terms of the *interest* we have in the connection between a voluntary action and its cited cause. He denies in this passage that such an interest is in a causal mechanism, for which a question in terms of 'How does it work?' is appropriate (p. 49). What he denies I would affirm with 'mechan-

---

[1] The phrase is used by Messrs. G. J. Warnock, P. F. Strawson, and J. F. Thomson, in their symposium in *Freedom and the Will*, ed. D. F. Pears, London, 1963.

ism' in scare-quotes. After all, there are interests and interests. Is it not *ever* the case that our interest in the connection between act and cause is precisely in learning *how* a man's habits or deliberate policies mesh with stimuli of certain kinds in yielding certain kinds of voluntary actions? My case for the affirmative must rest on my examples.

Donnellan has another very interesting argument against me. He concedes (1) that a person who causes another's voluntary action may be himself subject to blame for it by a third party, but claims (2) that in these cases the principal (second) voluntary actor cannot do the blaming. And yet (3) on my triggering model, it is "inexplicable why one should not blame another for triggering his mechanism or taking him over the threshold even though the act is voluntary" (p. 50).

I should like to concentrate my reply to this on the second statement and attempt to escape embarrassment by the time-honored tactic of making a number of distinctions. What are some of the more common ways in which one person can cause another to act voluntarily? First, he can provide an opportunity to the other to do what he was already bent on doing. Second, he can, like our fictitious bank teller, call the other's bluff or not heed his warning. Third, he can, like Mr. Black, direct the other's attention to something he might not otherwise have known. Fourth, he can exploit another's carefully studied habits, as in my "good morning" and "party bore" examples. Fifth, he can exploit the kind of "conscious habit" that is periodically subject to review, or a full-fledged deliberate *policy* of acting in certain ways in certain kinds of situations, as in my "ingenious suicide" example.

Now it would indeed be absurd for a voluntary actor to blame another for providing him with the opportunity for doing what he already had his heart set on doing. Donnellan asserts, similarly, that "it would be absurd for Mr. Green to blame Mr. Black for causing him to seek out Horner and kill him" (p. 50). This is not quite an example of "providing an opportunity," but it involves a similar principle, and Donnellan may be quite right in his judgment. (I am not sure he is right, however; I can imagine a heartbroken but unrepentant Mr. Green saying to Black, "Why did you have to tell me anyway? I'd have been better off deceived and unavenged than avenged and undeceived.") In the other three cases, however, I detect no impropriety, moral or conceptual, in the principal actor "accusing," "blaming," or "holding accountable" the one who caused him to act. There would be an absurdity in this only if the principal actor, in blaming the other, absolved himself; but I see no reason why, in these cases, he could not hold the other equally responsible with himself for what happened.

Mr. Lehrer complains, with some justice, I think, that in making my point against Hart and Honoré's voluntary intervention principle, I failed to consider their distinction between explanatory and attributive causal statements. Let me try to remedy that omission here, first by considering how Hart and Honoré's distinction is related to the causal distinctions that play a role in my paper. I used the term *causal citation* to refer to all causal statements of the form "Its cause was such-and-such." Some of these, by citing causal factors that are missing links in a person's understanding, function as explanations; but not all causal explanations are causal citations. An explanation of how a clock runs or how the tides come in, for example, can be what I called a "long-story explanation," and these in turn form a very diverse class indeed.[2] On the other hand, not all causal citations function primarily to explain. Prominent among the nonexplanatory causal citations, for example, are those made from what I called the "productive standpoint," where the concern is to cite the event immediately antecedent to a certain result that was sufficient in the circumstances for its production, whether or not that event contributes any light to a person seeking understanding.[3] Still other kinds of examples of nonexplanatory causal citations are found in my paper: citations of causal factors easiest to produce, eliminate, or control; or of greatest interest to a given profession; or of central concern for purposes of moral judg-

[2] The distinction between causal citations and long-story explanations is illustrated nicely by N. R. Hanson: "A farmer may ask what is causing his crop failure. But he will not, in the same tone of voice, ask what is causing its success. For he realizes that the success of crops depends upon a delicate atmospheric conspiracy of fine weather (but not too fine), warmth in moderation, rain but not flood, etc. Whereas, should his crop show signs of failing, he will hope that an expert will be able to say just what is the cause." "Causal Chains," *Mind*, LXIV, 1955, p. 310.

[3] Difficult as this distinction is to characterize in the abstract, it is intuitively clear, as is illustrated by the story of the straw that broke the camel's back. Suppose that Omar's daily routine is to take his camel to ten different stations in the marketplace. At each stop a merchant loads one hundred straws, one at a time, on the camel's back. By the end of the day, then, there are one thousand straws on the camel's back, a weight very near but not quite at the back's breaking point; and Omar then leads the camel to a warehouse to be unloaded. One day, just as the tenth merchant puts the last straw on the load, the camel sags to the ground, his back fractured. A subsequent inquest discloses that early in the day, at the third stop, a mischievous urchin, unnoticed by anyone, slipped one extra straw on the load. Now the question arises: Which straw caused the camel's back to break? Both the urchin's straw and the last straw (let us suppose) were necessary conditions of the break and, depending on what we include in the causal background, both can be regarded as sufficient. From the productive point of view, clearly the last straw was the cause, since *its* arrival at the top of the pile brought the burden above the breaking point. Still Omar does not *understand* why the camel's back broke until he learns of the urchin's straw, and the point of a causal *explanation* is to foster understanding.

ment; or enforced compensation or some other practical interest.

Hart and Honoré's distinction can easily be fitted into this scheme. In my terminology, it is a distinction between explanatory causal citations and one very special kind of nonexplanatory causal citation (*not* an exhaustive distinction between explanatory and *all* nonexplanatory citations). Hart and Honoré's "attributive" judgments always cite the actions of *human beings* as "the cause" of some *harm*. Attributive inquiries are often intertwined with, or dependent on, explanatory inquiries, but "searches for explanation are not the source of the lawyer's main perplexities: these arise when, after it is clearly understood how some harm happened, the courts have, because of the form of legal rules, to determine whether such harm can be attributed to the defendant's action as its consequence or whether he can properly be said to have caused it. These may be called attributive inquiries."[4] Whether or not a death is to be attributed to tuberculosis or to air pollution is in this special sense not an attributive inquiry; but, as the example in my article shows, citation of one or the other as "the cause" need not be explanatory either.

Insofar as I concerned myself with explanatory causal citations in arguing against the voluntary intervention principle, which was intended by Hart and Honoré to apply only to attributive causal statements (in their special sense), I was unfair to them, and Lehrer has a valid point against me. The point, however, is hardly as damaging as Lehrer contends, for my counterexamples seem to me to tell just as effectively against the voluntary intervention principle when it is understood to apply only to attributive causal statements. "It is quite clear," Lehrer writes, "that the death of Horner is to be attributed to the husband, Mr. Green, and to no one else. It surely would be a mistake to attribute his death to Mr. Black" (pp. 52–53). This is not at all clear or certain to me. I do not see how the facts of the case would logically foreclose the judgment that the cause of Horner's death was Black's spilling the beans or that the death is to be attributed jointly to Black and Green. The facts leave these possibilities open; our moral standards or legal rules or other practical purposes will determine which logically permissible attributive judgment we make.

"The voluntary intervention principle," Lehrer continues, "does not deny that Black's action might serve as a causal explanation of Horner's death; it only denies that Horner's death may be attributed to Black's action" (p. 53). Explanatory causal citations are relative to the gaps in a person's understanding; but I can imagine two persons who have no such puzzlement—they thoroughly understand exactly how Horner's death came about—and yet one "attributes" it to

---

[4] *Causation in the Law*, Oxford, 1959, pp. 22–23.

Black and the other to Green. Here there may be a harmony of understanding, and yet a clash of moral judgments or of appraisals of "expectability," or standards of normalcy or the like, leading to opposed attributive judgments.

I must hasten to add that I agree with Hart and Honoré that, after the facts are in, we do not have *complete license* to settle the attributive question in accordance with our moral standards, policies, and purposes, that certain common sense principles here place effective limits on our discretion. I disagree with them only in my account of what these principles are. I opt for the principle that "the more *expectable* is human behavior, whether voluntary or not, the less likely it is to negative causal connection." They prefer the voluntary intervention principle. The only way the issue can be argued is through examples and counterexamples.

Before leaving this question, I should like to make a stab (it can be no more than that) at locating the source of Lehrer's apparent certainty that the death of Horner is to be attributed to Green and no one else. Our disagreement on this question may be the result of an unnoticed ambiguity. Our language often provides us with alternative but equivalent ways of making causal judgments. We can use the language of causal citation or that of causal agency. Thus, we can say either that "Jones caused the door to close (by pushing it)," or "Jones closed the door." In many other cases, however, our language does not give us this option. I recently observed two young boys teasing a smaller child by causing him to giggle uncontrollably. Finally, one turned to the other and said (improperly): "We'd better quit or we'll laugh him to death." The boy wanted a transitive verb that could be used interchangeably with 'causing another to laugh'; but our language does not carry such a word in stock, there being little demand for it. In still other cases, transitive verbs of action are available to tempt us, but they cannot substitute for straightforward causal idioms without severe strain on sense. Thus, the ingenious suicide, in the earlier example, surely caused his own death (the cause of his death was his provocative remark to Manley Firmview); but we cannot express this by saying that he *killed* himself without being misleading in the extreme. The killing, Lehrer would quite rightly say, can be attributed to Firmview "and to no one else"; and the same point applies, *a fortiori*, to 'shooting' and other more determinate action verbs. Similarly, there might be a point, in some contexts, in saying that the cause of a certain married woman's pregnancy, after many unhappy barren years, was the medication prescribed for her by her obstetrician; but it would raise eyebrows to express this fact by saying that the doctor "got her pregnant," and it would be plain libel to say that he "fathered her child." This was done by the husband and "no one else." Thus the distinction between causing and doing can be of considerable moment.

Perhaps Lehrer would be willing to accept my counterexamples if I made it perfectly plain that I was restricting their application to attributions via causal citations, and making no further point about the language of agency.

The argument Lehrer borrows from Locke is a perfectly sound one, and its conclusion is important and insufficiently attended to by philosophers in our time. Its point is that it does not follow from the description of an action as voluntary (unconstrained, informed, and deliberate) that the agent could have done otherwise. If we use the term 'avoidable' for the more cumbersome phrase 'could have done otherwise,' the point can be put pithily: voluntariness does not entail avoidability. Furthermore, the compatibility of voluntariness with causal determination does not entail the compatibility of avoidability with causal determination. But while this is an important point (and one commonly obscured by the leaky-umbrella term 'freely'), it hardly counts against anything I have said, for in this paper, I defend a thesis about voluntariness only, and one that leaves the further vexatious question about avoidability and determinism entirely open.

EVALUATIVE METAPHYSICS*

NICHOLAS RESCHER

1. *"Taxonomic" v. "Evaluative" Metaphysics*

Metaphysics has from its very inception been conceived of as the study of being. On the one hand this was taken to involve "ontology": the analysis of the kinds or categories of things there are, undertaken with a view to a determination of the natural groupings involved, of their interrelationships, and of their most pervasive or most fundamental features. (On the older views, ontology endeavored to get directly at the actual structure of the world; more recently, i.e., since Kant, it has been the style to deal instead with the conceptual structure of our thought about the world.) For convenience, I shall refer to this structural or architectonic mode of metaphysics as taxonomic metaphysics (be it strictly ontological or conceptual taxonomy that is at issue).

One may contrast with *taxonomic* metaphysics a quite distinct mode of metaphysical inquiry for which I propose the characterization of *evaluative* (or one could perhaps say normative) metaphysics.[1] Simply put, its aim is not to *sort* (classify, categorize) but to *grade* (evaluate, rank). Its aim, to state it crudely, is not *to set up pigeonholes* for classifying "the furniture of the universe," but rather *to give it marks*, to evaluate it. The very possibility of such an enterprise rests on the case for the existence of distinctly "metaphysical" values—as opposed to ethical (right/wrong) or aesthetic (beautiful/ugly) or practical (useful/useless) ones. The presentation of this case is my objective here.

To borrow a distinction from P. F. Strawson, we may distinguish between the *descriptive* and the *revisionary* (or, I would suppose, *prescriptive*) mode of conducting metaphysical inquiry. "Descriptive metaphysics," writes Strawson, "is content to describe the actual

---

* I wish to thank the students in my recent Seminar in Metaphysics and my colleagues Kurt Baier and Jerome Schneewind for constructive criticism.

[1] These enterprises, albeit distinct, are clearly not independent. Indeed, it could be argued that taxonomic metaphysics is methodologically prior to evaluative metaphysics, since to evaluate what exists we must first classify it. Moreover, I have no intention of claiming these two to be the only modes of metaphysics. For then one very important omission would be *explanatory* metaphysics: the attempt to explain why certain pervasive features of the world are as they are and not otherwise (e.g., the dimensionality of the universe), or even to answer the question of why anything exists at all.

structure of our thought about the world, revisionary metaphysics is concerned to produce a better structure." (*Individuals*, London, 1956, p. 9.) It is clear that taxonomic metaphysics can, as Strawson shows, be pursued in either of these ways. It should be noted that Strawson's distinction between a "descriptive" and a "revisionary" mode of approach can be carried over intact to evaluative metaphysics. Here, too, one can either lay out the "value-structure" we actually employ in our thought about the world, or one can endeavor to produce a better one. Our present, rather illustrative than developmental, approach to evaluative metaphysics will confine itself pretty much to the descriptive side of the border.

2. *Historical Observations*

The paternity of evaluative metaphysics may unhesitatingly be laid at Aristotle's door. In the *Physics* and the *De Anima* we find him at work, not merely at classifying the kinds of things there are in the world, but in ranking and grading them in terms of relative evaluations. His preoccupation in the *Metaphysics* with the ranking schematism of prior/posterior—for which see especially Chapter 11 of Book 5 (Delta), and Chapter 8 of Book 9 (Theta)—is indicative of Aristotle's far-reaching concern with the evaluative dimension of metaphysical inquiry. It was a sound insight that led anti-Aristotelian writers of the Renaissance, and later pre-eminently Descartes and Spinoza, to attack the deeply Platonic/Aristotelian conception of the embodiment of value in nature. And it was a sound cognate insight that led the modern logical positivist opponents of metaphysics to attach the stigma of illegitimacy to all evaluative disciplines.

3. *Metaphysical Value*

We shall now attempt to establish the existence of metaphysical values and to clarify their nature by citing several examples. In each case, the value instanced is to conform to two essential characteristics of metaphysical valuation in that: (1) genuine valuation—i.e., some authentic concept of greater or lesser value—is at issue, and (2) the mode of value involved is *sui generis* in being genuinely metaphysical, and not ethical, aesthetic, pragmatic, etc.: it evaluates types of *things or conditions of things existing in nature* (not acts or artifacts) with a view to their intrinsic merit (not simply their "value-*for*" man or anything else).

4. *Exhibit A: The Value of Duration*

Our first example of an instance of metaphysical value is *duration*. The focal thesis here is that this is a metaphysical value, and that its status as such is exhibited by the following principle:

> Everything else being *equal*, that of two *kinds* of objects of the general type is better *sui generis* (i.e., *qua* examplar of this type) which lasts the longer, either collectively or with respect to its individual instances.

For the sake of an example, let it be supposed that we are given two kinds of mice, say A-mice and B-mice, such that while comparable in every other respect A-mice are longer-lived than B-mice. We are now entitled by the principle at issue to say, not that A-mice are *better* than B-mice, but that A-mice are *better mice* than B-mice.[2] The durability of a species, either collectively or distributively, constitutes (according to the principle) a *prima facie* point of merit, a positive value, in such a way that neither its numerosity nor spatial extensiveness (size) can afford any rival criterion of merit. (The question of why it does and they do not is, of course, a long story. It need not concern us here.)

5. *Exhibit B: The Value of Personhood*

A second example of an instance of metaphysical value is *personhood* or rather the cluster of functions and capacities that are involved here. I have in mind here such things as: life (i.e., self-maintenance and reproduction), sentience, locomotion, consciousness and thought, memory, will (i.e., the capacity for voluntary action), imagination (i.e., the capacity for hypothetical thought), conscience (i.e., a moral sense of right and wrong), and reason. With respect to these we have the following principle:

> Every capacity that is a characteristic constituent of personhood is of value, and the more intimately the capacity is inherently (or: *essentially*) characteristic of personality, the greater is its value (e.g., reason is of higher value than sentience).

A *locus classicus* of such evaluation is represented by Mill's well-known dictum that "It is better to be Socrates dissatisfied than to be a pig satisfied": the possession of "higher" capacities introduces a greater value over against the sole possession of "lower" ones—no matter how thoroughly they are present and how satisfactorily they are exercised. And Mill is surely not alone here: few men would hesitate to impute value to the constituents of personhood. Indeed, some metaphysicians have held that value can only exist in and apply to persons (e.g., McTaggart, *The Nature of Existence*, Secs. 788–791, and cf. 813).

---

[2] This position would seem to commit me to the view that U^{238} is better than U^{235} in view of its substantially larger life span. The escape hatch is, of course, provided by "other things being equal."

A profoundly metaphysical problem, to be sure one with deep moral ramifications, was confronted by a group of persons I read about in the newspaper some time back. This group was called on to develop criteria for selecting from among men suffering from some almost certainly fatal disease some few who could be given at one uniquely equipped hospital a fantastically complex course of treatment that is almost certain to effect a cure. For this group had to confront and to some extent resolve the question of the comparative value of variously constituted constellations of personhood-factors.

6. *Exhibit C: The Value of Knowledge*

A third example of an instance of metaphysical value is *knowledge*, not the value of possession of knowledge by individuals (however greatly this is to be prized), but the relative value of intellectual disciplines or branches of knowledge taken as abstract entities in their own right. We usually indicate such value by speaking of their "importance" and "interest." And in doing so we need not (although we, of course, may) intend these terms to be taken with a parochial significance ("interesting" = "interesting for *us*"; "important" = "*pragmatically* important for the attainment of practical life-aims"). The type of "interest" or "importance" at issue from a metaphysical standpoint is neither egocentric nor pragmatical, but is dictated by the *nature of the object*. The operative principle appears to go somewhat as follows:

> Of two branches of knowledge, that is the more valuable which, all else being equal, (1) deals with objects that are more extensive and comprehensive on the spatio-temporal scale, or (2) deals with objects more intimately relevant to (but not necessarily of any *practical* importance for) peculiarly human concerns.[3]

Of course these two criteria of *prima facie* value work against each other in theory as they do in practice.

The whole question of the relative value (sc. *intrinsic* value) of branches of knowledge and disciplines of learning exhibits the characteristic marks of metaphysical valuation. I need not elaborate here upon the great practical urgency of this problem in a day and age in which intellectual disciplines must compete with one another for funds, for recruits, for their "place in the sun" of the academic curriculum, for library space, and the like.

---

[3] Of course, the group of objects dealt with is only one consideration among others. Astronomy and astrology deal in large measure with the same objects, but clearly "other things" are not equal.

When I contend (as I am prepared to contend) that *Egyptology is of value*, I neither say nor imply either that I am interested in this subject or that the material well-being of the race would be adversely influenced by its neglect. When I contend (as I am prepared to contend) that *physical cosmology is more important than Egyptology*, the issue certainly *does not* turn on my or anybody else's personal predilections. It certainly *does* turn in part on the fact that one discipline focuses upon a unique, restricted episode of the past (albeit of the *human* past), while the purview of the other spans awesomely vast reaches of the future. But the matter of *practical* relevancy does not arise (or may be supposed not to do so).

We thus conclude our list of three "exhibits" of items that could, on "the ordinary view of things," be regarded as possessing value of the metaphysical type, viz., duration, personality, and knowledge.

In everyday-life economics, once a kind of thing is fixed upon (so that only *sui generis* comparisons are in question), we conceive the principle determinants of value to be *quality* and *quantity*. In the metaphysical valuation of natural kinds, duration (but neither spatial extensiveness nor numerosity!) is the counterpart of quantity, and the complex of personhood characteristics provides a counterpart of quality for the animate realm. (With inorganic nature, and with such nonspatio-temporal things as branches of knowledge, the situation is more complicated.)

7. *The Appropriateness of Valuation*

A number of rather dogmatically enunciated valuations have been put forward in the "exhibits" of the three preceding sections. It is possible, indeed highly probable, that the correctness of these specific valuations may be denied by many. But any denial of their correctness which merely plumps for some rival alternative valuation does not go against the fundamental position being advanced—viz., that it is possible to apply the mechanisms of valuation to such matters as duration, personality, and knowledge, and to do this in such a way that neither ethical nor aesthetic nor practical (etc.) values are at issue. To the rival value-metaphysician with an alternative schedule of valuation we can extend a hand of welcome into the lists of battle, for he is ready to "play our game." The only real opponent at this level is he who denies root and branch the applicability of the value-concept throughout this entire domain of things that lie wholly outside the area of the "practical politics" of human action. To such an opponent one may proffer the balm of sympathy, for his is no easy row to hoe. And in any case, since the value outlook of our ordinary life patently goes directly counter to his views, the burden of proof lies with his position.

8. Characteristics of Metaphysical Values

The several types of metaphysical value which have been instanced possess various common characteristics:

1. They are all *values*, i.e., instances of the application of such evaluations as better-or-worse and more-significant-or-less-significant.
2. They are *"metaphysical"* values in that they relate to the *intrinsic* merit of existing things, their "desert for existence." Such values are neither aesthetic (having to do with "enjoyment" in contemplation, primarily in respect to artifacts), nor ethical (bearing *directly* upon the evaluation of human acts), nor pragmatic (relating to use or consumption or "enjoyment"). They relate to the *being of things*, not necessarily to the realm of *human* purposes or interests.

Being *sui generis*, metaphysical values exhibit the extension of the schematism of values beyond their most common and familiar range.

9. Metaphysical Value and the "Ethics" of Creation

Although we have been concerned to maintain that metaphysical values are not ethical, it should be recognized that, from a somewhat far-fetched and certainly extraordinary perspective, they can be looked upon as being "ethical" in a rather extended sense. Let us envisage the project of ethical appraisal of the creation-choices of a hypothetical creator-deity, whom we may assume to be omnipotent and omniscient. That is, let us perform the thought-experiment of envisaging alternative hypothetical possible universes and then undertake the ethico-moral appraisal of a postulated creator-deity who would select *this* possible universe for actual realization (creation). We come to the core of the question regarding metaphysical values ("Is it better that X's should exist in a world or not, other things being equal?") through the instrumentality of the corresponding (quasi-ethical) question regarding world-creation. ("Is greater ethico-moral merit to be accorded to a deity who creates a world that contains X's rather than not, other things being equal?") Metaphysical values, although not themselves of a character appropriately subject to ethical considerations of the ordinary kind, can be brought within an extended sphere of ethical evaluation via the hypothesis of a deity who chooses to *create* or *omit* or to *continue* or *annihilate* (destroy) the items whose value is at issue.

An objector might argue: "As long as we must do ethics in conjunction with doing evaluative metaphysics, we may as well take care of questions about the metaphysical implications of moral judgments within our ethical system." One of my (ethicist) colleagues, indeed, was unkind enough to speak of "gerrymandering" in connection with the notion of evaluative metaphysics. I think this objection is a two-edged sword. As long as it is supposed that ethical evaluation is directed at *human* acts (surely a plausible supposition!) and also that metaphysical evaluation is directed in the main at *objects that lie outside the sphere of possible human action*, it hardly seems plausible to class evaluative metaphysics as a branch of ethics. But there is, admittedly, a considerable area of close contiguity. Consider an example: A given person has an existing capacity for, say, aesthetic appreciation or the exercise of intelligence. It is a thesis of human ethics that a person should exercise this capacity to the highest possible extent. The cognate thesis of "creation-ethics" is that the postulated creator should set this capacity at the highest possible level. (Note the connection between these two theses: it would clearly be paradoxical to hold that, while it is a good thing that a person should exercise an existing capacity to the highest possible extent, it would not be a good thing if this capacity were to be augmented.)

The classical instance of this general mode of ontological valuation is, of course, afforded by Leibniz who has a well-defined formula for the metaphysical value ("desert" for creation) of possible worlds, namely, "the variety of phenomena" taken together with "the simplicity of laws."

A more modern example of this line of methodology in recent philosophy is G. E. Moore's *Principia Ethica*. Moore's celebrated "method of absolute isolation" (Secs. 50, 55, 57, 112–113) invites us to make comparative evaluations of two hypothetical worlds supposed to be alike in all relevant respects except that in one of them some factor is exhibited which is lacking in the other. Thus Moore argues for the intrinsic value of natural beauty (i.e., its value even apart from human contemplation) by the argument:

> [A hypothetical] beautiful world would be better still, if there were human beings in it to contemplate and enjoy its beauty. But that admission makes nothing against my point. If it be once admitted that the beautiful world *in itself* is better than the ugly, then it follows, that however many beings may enjoy it, and however much better their enjoyment may be than it is itself, yet its mere existence adds something to the goodness of the whole: it is not only a means to our end, but also itself a part thereof [*Op. cit.*, Sec. 50].

To espouse the project of evaluative metaphysics is to side with Moore against Sidgwick's thesis that: "If we consider carefully such permanent results as are commonly judged to be good, other than qualities of human beings, we can find nothing that, on reflection, appears to possess this quality of goodness out of relation to human existence, or at least to some [presumably animal] consciousness or feeling" (*Methods of Ethics*, I, ix, Sec. 4). (There is, of course, the trivial fact that if "we" do the considering, "we" do the evaluating. The point to be borne in mind is that this need not be done from "our" standpoint.)

Moore was clearly aware of the distinctiveness-in-the-face-of-kinship between standard ethics on the one hand and evaluative metaphysics on the other, recognizing the *sui generis* character of the latter enterprise:

> By combining the results of Ethics as to what would be good or bad, with the conclusions of metaphysics, as to what kinds of things there are in the Universe, we get a means of answering the question whether the Universe is, on the whole, good or bad, and how good or bad, compared with what it might be: a sort of question which has in fact been much discussed by many philosophers. [*Some Main Problems of Philosophy*, p. 40 of the Collier paperback edition.]

Our suggested method goes beyond Moore, although not Leibniz, in realizing the possibility of transforming such a strictly metaphysical evaluation into the quasi-ethical one through the means of a "creator hypothesis," i.e., the assumption of an (omniscient and omnipotent) deity who acts to actualize the hypothetical world in question, and whose assumed creation-act is, like other acts of persons, capable of moral evaluation.

10. *Points of View Upon "Creation Ethics"*

The possibility of a divergence between human and divine standards of valuation in the assessment of divine acts of creation/omission and continuance/annihilation has long been recognized. It is inherent in the biblical story of Abraham's plea for Sodom, when the patriarch "drew near [to the Lord] and said, Wilt thou also destroy the righteous with the wicked?" (*Genesis* 18:23)

From the human, or perhaps better *humane*, standpoint the pivotal consideration in evaluating possible worlds relates to the presence in a universe of such matters as pain, anguish, suffering, anxiety, and everything else relating to physical or mental discomfort or distress and their opposite benefits. The relevant standard of valuation is simply this: the less of this sort of thing is contained in a possible world, the better; and the more, the worse (other things being as equal as possible, in particular with respect to the presence and functioning of sentient and rational creatures).

From the divine, or perhaps *superhuman*, standpoint the kinds of pleasures and discomforts relating to sentience (actual or anticipated) need not necessarily be accorded predominant, or even great (if indeed it can have any) weight. The purely rational *intellectual* merits exhibited by possible worlds are now the pivotal considerations. We have here to deal with the purely rationally determinable characteristics of the universe: its symmetry v. asymmetry, order v. chaos, richness v. aridity, variety (diversity) v. uniformity (sameness), etc., i.e., only those features of the world which conduce to its capacity to give satisfaction to a comprehending intelligence. Spinoza (*Ethics*, IV, 26) says:

> *Mens, quatenus ratione utitur . . . [nihil] sibi utile esse judicat, nisi id, quod ad intelligendum conducit.* ["The mind, insofar as it uses reason, adjudges nothing as profitable to itself excepting that which conduces to understanding."]

According to Leibniz, when the deity contemplates the various possible worlds *sub ratione possibilitatis* prior to the creation in order to determine that which is the best, the criterion of merit he applies is to find that which exhibits the greatest "variety of phenomena" embraced in the world consonant with the "simplicity of the laws" governing the interrelation of its occurrences.[4]

It is of course possible—and indeed perhaps even necessary—to conceive of our postulated deity, hedged on all sides by *omni-s*, to confine himself to what we have here called the *superhuman* point of view, wholly to exclusion of the humane. This, then, would be the "God of the Philosophers" who, irrespective of how his creatures may love *Him*, can at best respond with an *amor intellectualis*. To quote Spinoza again (*Ethics*, V, 36):

> *Mentis Amor intellectualis erga Deum est ipse Dei Amor, quo Deus se ipsum amat, non quatenus infinitus est; sed quantenus per essentiam humanae Mentis, sub specie aeternitatis consideratam, explicari potest, hoc est, Mentis erga Deum Amor intellectualis pars est infiniti amoris, quo Deus se ipsum amat.* ("The intellectual love of the mind toward God is the same as the love of God itself with which He loves Himself, not insofar as infinite but insofar as He can be manifested through the essence of the human mind considered *sub specie aeternitatis*, that is to say, that the intellectual love of the mind toward God is part of the infinite love with which God loves himself.")

---

[4] J. M. E. McTaggart (*The Nature of Existence*, Sec. 813) in attempting a complete list of "qualities which are good" arrives at: knowledge, virtue, "the possession of certain emotions," pleasure, "the amount and intensity of consciousness which we may call 'fullness of life,' " and "harmony." He accepts all these values except for the last "because I see no good or evil under this head which does not come under the other five." Conveniently for his thesis (that value can apply only to selves), he overlooks the fact that this value alone can obtain in a universe in which there is no organic life (and so no persons or selves).

It is important for present purposes to bear in mind this distinction between how we conceive a creator-deity to deploy a criterion of merit in evaluating possible worlds and how we humans would go about it ourselves.

This distinction in question is important for present purposes because it is helpful for making a point which seems to me correct, viz., that the evaluative "point of view" from which a supposed creation choice of a hypothetical creator-deity is assessed can be avowedly based upon *human* criteria of valuation and need not be of deo-centric character. Evaluative metaphysics is not a branch of speculative theology. It is not a part of theology in *any* shape or form. It is rather (to make a parody of it) the project of making a reasoned and systematic projection of human valuations onto a cosmic screen.

11. *A Foothold for the Method*

Finally, I turn briefly from description to argument. I shall endeavor to argue that ethical values in the strict sense are in the final analysis "grounded in" or "bound up with" or perhaps even "based upon" metaphysical ones. By an "ethical value in the strict sense" I mean one that relates to evaluative assessment of the moral rightness or wrongness of human action or behavior. It is my thesis that the ethical evaluation of human actions is inseparably bound up with the metaphysical evaluation of factors extending beyond the domain of human action; this being so in the precise sense that one could not reasonably maintain the ethical value without being prepared to support the metaphysical one. It is my contention that ethical valuation involves a commitment to evaluations of the metaphysical type.

The argument, in its abstract generality, runs as follows: Consider any strictly ethical thesis such as, say, "doing X is wrong"—e.g., "the infliction of needless pain is wrong." It is clear that any type of human action such as "doing X" has a "strictly ontological counterpart" in the types of objective, de-agentized situation which come to be realized once such an action is supposed to have been performed. (I think here, not of "consequences," but of *the situation itself* taken in abstraction from the agent who brought it about.) Thus "X has inflicted needless pain upon Y" leads to "Y is undergoing a needless pain" and "W has murdered Z" leads to "Z's life has been prematurely terminated," etc. The performance of any ethically wrong act creates a "situation" for the affected persons and/or for the perpetrating agent himself, a situation which must, consonant with the classing of the act as (morally) wrong, be itself characterized as (ontologically) bad. For the ethical evaluation (as right or wrong) of a type of human action involves (commits one to; presupposes) a certain metaphysical evaluation of its strictly ontological resultant counterpart situation in such a way that the former simply could not be maintained uneccentrically (i.e., other than unreasonably) without acceptance of the latter. That is, anyone who wants to hold that "infliction of pain is (other things being equal) wrong" must be prepared to hold that *the undergoing of pain* is (other things being equal) bad; and anyone who wants to hold that "deprivation of life is (other things being equal) wrong" must be prepared to hold that the *maintenance of life* is (other things being equal) good; etc. We thus submit that ethical values in the final analysis "spill over into" metaphysical ones because the espousal of ethical valuation of certain human choices of courses of action calls for, and is inextricably bound up with, cognate corresponding valuations of the metaphysical type. The ethical evaluation of human acts cannot in the final analysis be carried through without the evaluative assessment of items whose presence or absence does not turn exclusively upon human endeavor and whose evaluation, therefore, is not of a purely ethical character.

Espousal of any ethical scheme for the moral evaluation of human behavior carries with it an implicit commitment to an underlying value-metaphysic on which the ethical valuations rest for their support. In approving and disapproving of various types of human actions we are engaging in a kind of "ethics of creation" albeit on a very microscopic scale: we are committing ourselves to various theses about what at any rate one very small part of the world (i.e., that within the sphere of human control) ought to be like. In this way the analysis of the foundations of ethical valuations comes to be seen as an important beginning task, although perhaps by no means the only proper work, of evaluative metaphysics.

12. *Summary*

In this paper I have tried to do primarily three things:

1. To show, primarily by examples, that there is in addition to taxonomic or classificatory metaphysics another, distinct, metaphysical enterprise whose aim is to evaluate or rank.
2. To show that such a form of metaphysics is bound up with any application of ethical considerations at the theological level through the quasi-ethical evaluation of hypothetical creation-acts.
3. To show that the espousal of any (ordinary) ethical code involves an at least partial, implicit commitment to an evaluative metaphysic.

COMMENTS

LEWIS WHITE BECK

For there to be an evaluative metaphysics, three conditions must be met: (a) there must be distinctively and irreducibly metaphysical values; (b) there must be a valid argument that (a) is true; and (c) there must be a procedure for deciding what manifests these values. I wish to comment on the first and third of these conditions.

(a') Mr. Rescher proposes three instances of metaphysical value, out of a vast number that have been propounded by metaphysicians in the tradition of the Great Chain of Being. The first he proposes, duration, seems to me to be a measure of the quantity of being itself, with being the root value as opposed to nonbeing; since Mr. Rescher considers duration as the axiological dimension to quantity, I do not believe he would disagree with this interpretation. The third of his metaphysical values, knowledge, seems to me to be a special case of his second. Hence, he seems to me to say: personal qualities and personal beings are better than nonpersonal; the more personal the better; and the longer-lasting the better. One exemplification of metaphysical value Mr. Rescher needs in the later part of his paper, however, is missing from his formal list, viz., absence of pain. He later holds that freedom from pain is a positive metaphysical value, but it does not seem to me to be more intimately connected with personhood than its opposite. I shall return to this point in a moment.

(c') Metaphysical values are assigned in a thought-experiment, using the fiction of an omnipotent moral agent, G. G is like a human moral agent except that G has the power to produce infallibly some X which a human being could not produce, or could not infallibly produce. If I judge G is praiseworthy if G produces X, this is the necessary and sufficient condition for saying that X has positive metaphysical value.

Now what is accomplished by the ascription of positive metaphysical value to X? Mr. Rescher holds that I need to say X has positive metaphysical value if I am uneccentrically to hold that it is right for a human being to try to produce X. 'You ought to produce X' is morally correct if and only if the corresponding ontological situation, "X" or "your producing X" (I cannot be sure here which Mr. Rescher means) is metaphysically better than its alternative.

I cannot hold, according to this argument, that your producing X is morally valuable "in a strict sense" unless I hold that the situation of X's being produced is metaphysically valuable; and I cannot hold it to be metaphysically valuable without deciding that I would give G a good mark if G produced X. But the criteria for deciding whether to give G a good mark for producing a world containing X are exactly the same as those I have for giving *you* a good mark if you try to produce (or exemplify) X. And as Mill saw, in one of his purplest passages, the difference in metaphysical status between G and man has no bearing on the values exemplified. If I have to know that X is good if I know that it is right to try to produce or exemplify X, as Mr. Rescher holds, than I have to know that X is good before praising G for producing a world that contains it. Hence, the thought-experiment contributes nothing to the decision problem.

Why, then, does he want to take this detour into metaphysics? I conjecture the following steps in his thought. (a) There are two kinds of judgments in ethics, one of which is ethical "in a strict sense." (b) They are so different, however, that it is permissable to assign one to one discipline, and the other to another. Hence, non-strictly ethical values become one species of metaphysical values. (c) But whereas the decision procedure in ethics is often clear or at least familiar to us, that in metaphysics is not; hence, we employ a quasi-ethical procedure in metaphysics, straining out everything peculiarly or strictly ethical in the evaluations ventured upon. (d) Whereupon we return to the ethical in a strictly ethical sense, having picked up a metaphysical value along the way which gives the ethical judgment some metaphysical anchorage.

All this is plausible, and all of it is familiar except the words "metaphysical value." For "metaphysical value" is, in the context of the decision procedure, the well-known ethical value commonly called "the good." Mr. Rescher's dialectic of ethical and metaphysical values is the well-known dialectic of the right and the good, somewhat oversimplified.

My evidence for this conclusion is twofold. First, the "metaphysical value" which functions most prominently in his actual decision procedure is "absence of pain," which is a prime example of the "good" as it has functioned in ethics, but which is not one of the metaphysical values mentioned in the beginning of his paper before he became involved in the quasi-ethical decision process. Second, the metaphysical values established by a Gedankenexperiment very unlike the thought-experiments we repeatedly make within ethics itself, seem, to that degree of difference, not to be well founded. For instance, the second and third exemplifications of metaphysical value seem to me to be well established precisely because we have an ethical occasion and ethical apparatus for establishing the values of personhood; and, on the other hand, the first of his metaphysical values is not established (at least to my satisfaction) just because we do not have an ethical occasion and apparatus for evaluating *duration sui generis*.

Mr. Rescher speaks ironically of making a parody of evaluative metaphysics by describing it as "the project of making a reasoned and systematic projection of human evaluations on a cosmic screen." I see nothing of parody in this statement; it seems to me to describe very well Mr. Rescher's project. But I would like to know why this project, or this projection, is undertaken. What value questions does this cosmic projection illuminate? Why do metaphysics unless we have to?

COMMENTS

THOMAS E. PATTON

Professor Rescher tries to show that "metaphysical valuation," as he calls it, is a game that we can play, losing ourselves, while millions in a cruel world hunger for a main course, in the problem of "desert for existence." This is said to be "genuine valuation" and to involve "some authentic concept of greater or lesser value." Central to it are what he calls "metaphysical values," *sui generis* phenomena, "neither aesthetic . . . nor ethical . . . nor pragmatic," which relate "to the *intrinsic* merit of existing things" and "to the *being of things*, not to the realm of human purposes or interests."

Rescher argues in terms of three examples, in which we see both particular "metaphysical valuations" and certain principles on which they are based. He concludes from these "that it is possible to apply the mechanisms of valuation to such matters as duration, personality, and knowledge." But what are the "mechanisms of valuation"? Much has been said on this topic in recent decades, too much for me even to summarize now, but let me just mention the central theme that "genuine valuation," as opposed, say, to the expression or evocation of feelings, involves the notion of *justification* for utterances of the form 'X is good,' for example. I would not be surprised if this justification had in all cases to lead back finally to "the realm of human purposes or interests," but I will not urge anything so dramatic here. For present purposes it is enough to remark that when 'X is good' is justified by appeal to a *principle*, as happens in Rescher's examples, it is in standard cases a principle which people in the plural in fact hold. Rescher's principles, however, in his three examples, seem conspicuously to lack the kind of social existence that "genuine valuation" is widely felt to demand.

Of course people might in fact adopt such principles, thereby making "metaphysical valuation" possible. An overwhelming question then arises: Why would they? In regard to moral discourse, one can at least say "These language-games are played." In regard to Rescher's project, can one not at most ask "Wouldn't this be an idle game?"

Never doubting, however, that evaluative metaphysics has a leg to stand on, and perhaps believing too in footholds for the effete, Rescher seeks finally to provide one for the method. Thus he argues "that ethical values in the strict sense are in the final analysis 'grounded in' or 'based on' metaphysical ones." But all that he in fact provides, I fear, is a prize specimen of specious reasoning.

Its primary claim is that "the espousal of ethical valuations calls for, and is inextricably bound up with, cognate corresponding ones of the metaphysical type." This depends on Rescher's "strict sense" for ethical values, hinted at before and stated explicitly now, which limits ethical valuation to *"human* action and behavior." It also depends on the doctrine that to every right or wrong action there corresponds a *"resultant* situation" that is respectively good or bad. But the situation cannot be *morally* good or bad, cannot have *ethical* value, being neither an action nor behavior. By a curious leap from this, evidently, Rescher concludes that the situation is "ontologically" good or bad, that what it has is *metaphysical* value.

We need not even ask here whether Rescher's "strict sense" for ethical values is too narrow to be acceptable philosophical jargon. For when Cain kills Abel, when Iago tricks Othello, when Communist agents plot the undoing of Daddy Warbucks, if there are situations that must be called bad *consonant* with calling these acts morally wrong, then is it not bizarre to suppose that the badness of these situations has nothing to do with "the realm of human purposes or interests"? Surely it must have everything to do with the ethical criteria that qualify these acts as wrong. Rescher's *non sequitur* that it is *metaphysical* badness then seems impossible to repair.

His argument may also fail in that he misuses the expressions 'based on,' 'grounded in,' and 'rests on.' This issue is obscured by what I take to be his less than precise description of examples. Thus in one case, explaining how situations correspond to actions, Rescher says that " 'X has inflicted needless pain upon Y' leads to 'Y is undergoing a needless pain' " and then that "anyone who wants to hold that 'infliction of pain is . . . wrong' must be prepared to hold that *the undergoing of pain* is . . . bad." This last is hardly plausible, however, for why must such a one hold any more than that the undergoing of *inflicted* pain is bad? The "resultant situation" here seems to have been described too broadly for its "metaphysical value" to run properly parallel to the ethical value of the action.

I think that this line of criticism, pursued far enough, would oblige Rescher to adopt descriptions of situations that correspond in all details to those of their counterpart actions. Thus in his example, the "situation which comes to be realized" on performance of the action mentioned in 'X has inflicted needless pain upon Y' must be characterized not by 'Y is undergoing a needless pain' but by 'Needless pain has been inflicted upon Y by X.' Similarly, the "strictly ontological counterpart" for the other example, in which 'W has murdered Z,' will have to be indicated, not by 'Z's life has been prematurely terminated,' but by 'Z has been murdered by W.'

If Rescher's counterpart situations are to be so redescribed, however, it becomes dubious that ethical values are "based on" or "grounded in" or "rest on" metaphysical ones in any non-Pickwickian senses. I think that the following parody is not too bad: "On what do you base your claim that John hit Mary?" "On my knowledge that Mary was hit by John." A theorist bent on making ontic hay in the light of such examples might discover that for many "active facts" there is a counterpart "passive fact" such that knowing the one is "inseparably bound up with" knowing the other. Harder heads among us might object that no "basis" has yet been given for the quoted claim and that the "two facts" are "inseparable" simply in virtue of being really one.

I suspect that some such objection may apply to what Rescher says about actions and situations and their respective ethical and metaphysical values. In the senses that he must intend, for example, I suspect that to call an action wrong will be much the same as to call the counterpart situation bad, so that the one judgment cannot happily be held to be "based on" or "grounded in" or to "rest on" the other. Of course to call an action wrong is not *just* the same as to call a situation bad, if only because an action is not a situation. But is it *just* the same to know that John hit Mary and to know that Mary was hit by John? One might say not, for what someone did is not the same as what happened to someone. Yet this, while perhaps true enough, leaves the lesson of the parody unimpaired.

REJOINDERS

NICHOLAS RESCHER

The general tendency of both of my critics—and also of some others with whom I have discussed the problems here at issue—is to consider metaphysical values as being merely crypto-ethical. Thus Mr. Beck writes: "The criteria for deciding whether to give G a good mark for producing a world containing X are *exactly the same* as those I have for giving you a good mark if you try to produce X" (p. 74). And Mr. Patton writes: "Rescher's principles, however, in his three examples, seem conspicuously to lack the kind of social existence that 'genuine valuation' demands" (p. 76). But Patton's stricture simply prejudges the point at issue, and Beck's criticism seems to me implausible. If *you* try, say, to produce a world in which the natural laws exhibit great symmetry and simplicity (for example) you would qualify for an insane asylum; G, however, fares differently. The objects of metaphysical valuation do not, in general, qualify as objectives of human action at all, lying wholly outside the realm of human practicabilities. There are, to be sure, important linkages between ethical and metaphysical evaluations—as exemplified in one instance by the kinship between the maximum *exercise* of an *existing* capacity on the one hand and the maximum level of *endowment* of a *potential* capacity on the other. Such linkages are a subject for analysis and study and are not to be settled by some sweeping, omnibus move, either of dismissal or of equation.

Mr. Patton finds difficulties in the notion of a "resultant situation" ensuing upon performance of actions, perhaps justly so since I could well have developed the point more fully than I did in the paper and, indeed, more fully than I can do now. But let me at least indicate the general orientation of my position. A reasonable paradigm for a broad class of ethical judgments is:

It was wrong of X to do A to Y (or perhaps: It was wrong of X to do A, since this affected the interests of Y adversely).[1]

Schematically we might represent such a judgment as:

(1) Wrong (bad) that: X do-A-to-Y.

Now Patton appears to think that by the "situation" resulting from the action of (1), viz., "X do-A-to Y," I mean simply the referent of

[1] And of course, correspondingly on the positive side: It was right of X to do A for Y (or perhaps: It was right of X to do A since this affected the interests of Y favorably).

its passive re-rendering, "Y be-done-A-to-by X," and that, while I would characterize (1) as "ethical," I would characterize as "metaphysical"

(2) Wrong (bad) that: Y be-done-A-to-by X.

But this is simply wide of the mark. The metaphysicalized cognate I have in mind for (1) is not (2)—certainly not!—but rather a depersonalized (or rather de-agentized) and objectified counterpart of (1), viz. something like:

(3) Wrong (bad) that: Y should undergo being-done-A-to.[2]

And even this is in fact not quite sufficiently de-agentized to take it out of the ethical and into the metaphysical arena and should be changed to read:

(4) Wrong (bad) that: Y should undergo what he is put into the position of undergoing as a result of being-done-A-to.

The emphasis should be on the paradigm: 'It is wrong$_1$ for X to inflict needless pain upon Y because it is wrong$_2$ that Y should undergo needless pain.'[3] The point of my argument is twofold: (1) that 'wrong$_1$' represents an ethical valuation, whereas 'wrong$_2$' represents a metaphysical one, and (2) that this sort of linkage between ethical and metaphysical valuations, with each type "spilling over" into the other—but neither strictly "reducible to" the other—represents a reasonably standard state of affairs.

Thus when Mr. Patton writes, "I suspect that to call an action wrong will be much the same as to call the counterpart situation bad," he attributes to me a view I cannot but reject as grossly oversimplified. The rightness or wrongness of actions is determined on the basis of standardly ethical considerations. The goodness and badness of counterpart situations I take to rest upon metaphysical considerations. The linkages between these two groups of considerations seem to me to be complex, interesting, and worthy of analysis. I hold no brief for any simple omnibus solution let alone an equation of the two groups of issues. I am sure a great deal more than I have been able to say here can and needs to be said on this head.

Mr. Beck asks: "Why do [evaluative] metaphysics unless we have to?" And Mr. Patton displays much the same reluctance when he writes: "Of course people might in fact adopt such [evaluative]

---

[2] Or the equivalent counterpart of (2): Wrong (bad) that: Y be-done-A-to.

[3] What is to be characterized as wrong$_2$ here is—*pace* Professor Patton—not the undergoing of *inflicted* needless pain, but the undergoing of needless pain *simpliciter*. That which is not of itself bad (or good) cannot, surely cannot, become so simply by the productive intervention of human agency.

principles thereby making 'metaphysical valuation' possible. An overwhelming question then arises: Why would they?" Both critics seem to have displayed a marked disinclination to the enterprise. But in doing so they seem to me to take, against what I am sure is their own considered view, a patently unphilosophical position. After all, why should we *need* or *have to* pursue any branch of philosophy, or indeed any other branch of intellectual inquiry: Can we not appropriately do so largely on the basis of "intrinsic" interest? More crucial, perhaps, is our discipline's long-standing interest in reasoned *systematization*. The philosopher cannot but be interested in the concept of value, in the range of its application, and in the nature of its application throughout that range. Once it is granted that there is scope for metaphysical valuation—and both critics seem to grant this, albeit but grudgingly—then the question "Why concern yourself with the metaphysical applications of value?" ought to elicit the reply: *Because they are there.*

THINGS AND QUALITIES

HERBERT HOCHBERG

A white square patch is a simple or bare particular exemplifying the universals or qualities white and square. A white square is a combination of qualities. A white square patch is a combination of instances of qualities. These three statements sum up proposed ontological analyses of things and facts centering around the problems of universals and individuation. Without making at all clear what 'to analyze' means in this context, I propose to consider these views. In particular I wish to re-examine arguments directed against the second on behalf of the first. What I hope to show is that these arguments are ambiguous, mistaken or question begging, in short, bad. What is to be suggested is that the second view is not as unacceptable as some would have us think. In connection with these questions about "what a thing is" we shall confront a further question about the distinction between numerical and conceptual difference. What is further hoped is that the comparison of our three alternatives will shed light on related issues regarding identity and difference. While explicit references will be minimized, the paper owes much in its concern with issues, and for arguments that it purports to refute, to early papers of Moore and Russell and to the writings of Gustav Bergmann. In the attempt to defend the second alternative, a taint of Bradley may be detected. I mention these names also in lieu of providing in this paper a consideration of what sort of thing one is getting at in any of the above "analyses" or "ontologies."

The paper is about two white squares. Forgive the barbarism that reading imposes if I refer to them as 'Big Socrates' and 'Big Plato.' The sentences 'Big Socrates is white,' 'Big Socrates is square,' 'Big Plato is white,' and 'Big Plato is square' describe, or are about, the two objects. Or, one might prefer to say, they assert facts about these objects. In virtue of these facts the sentences are true. The nonexistence of these facts would make these sentences false. In view of this some say that the sentences refer to facts when true. Some would further have it that these facts are composed of a particular and a universal in the ontological tie of predication.

On this view the "ordinary" objects that we started talking about, Big Socrates and Big Plato, are either thought of as *composed* of particulars exemplifying a set of qualities and, hence, as a sort of conjunction of facts, or as being *dissolved* into these components. That is, they are either composites or nonexistent or, perhaps, both.

For one might hold that only the simple elements "exist" in a basic sense of that term while composites do not. On the other hand, Big Socrates differs from Pegasus with respect to some sense of the term 'exist.' Hence, one might hold that he exists in some sense but not as an element in one's ontology. Relating him to the ontological view we are sketching, he would have to be a composite of the kind mentioned above, a sort of conjunction of facts. However, a proponent of this view might insist that this way of putting the matter is misleading. For, in an "improved" language which would more clearly reflect his ontology, no term would occur which would refer to Big Socrates. What would occur would be a term referring to the bare particular that enters into the facts, which we might commonly think are about the composite patch. No such particular would "correspond" to Pegasus. The point would be that, because Big Socrates is a composite, the term 'Big Socrates' would not be a name. Hence, that Big Socrates is a composite is shown by the analysis on this view and not stated after it is made. Perhaps, however, a proponent of this view would accept the identification of Big Socrates with a conjunction of facts. In any case, I put the matter this way since the question of whether or not the term 'Big Socrates' is the name of a composite of some kind is a point that will be at issue and hence not to be presupposed.

On this view, then, what is named by a proper name in the case of Big Socrates is the bare particular that is present in all facts about the white square. Thus, while in one sense Big Socrates is identified with a set of facts, in another he is identified with a component of those facts. Let us call these peculiar components of each of our white squares Socrates and Plato, respectively. Big Socrates is, then, what we started out with, and Socrates is what metaphysics has reduced him to. These special constituents of our white patches or, if you prefer, bare particulars, are also thought to account for the numerical difference between the two patches. Irrespective of any qualitative difference between Big Socrates and Big Plato they are different in virtue of having different bare particulars, Socrates and Plato. These latter are simply different. Note that a bare particular is such that to speak of no qualitative difference between Big Socrates and Big Plato is to speak of no qualitative difference between Socrates and Plato. For, on the bare particular analysis, it is the bare particulars which exemplify qualities or universals. Just as bare particulars are thought to account for numerical difference, universals are the ground of sameness, or perhaps better, universals account for both Big Socrates (Socrates) and Big Plato (Plato) being the same in some respect. The ground of difference and the ground of conceptual identity are thus joined by predication to constitute facts.

In such an ontology we then have (a) bare particulars, (b)

universals, (c) the tie of predication, and possibly (d) facts. I say "possibly" facts in view of the question that was raised above about composite entities, for facts are composites. But as this may be thought to simply legislate about the term 'entity,' or 'existent,' we can, if need be, talk about different senses of that term. This would acknowledge that facts do not exist in the sense in which bare particulars and universals do, i.e., facts are not simples. Once facts like *that Socrates is white* are acknowledged as existents, in some sense, Big Socrates may be considered to exist in a sense other than that in which Socrates does: as the facts about Socrates. Hence, we may add (e) conjunctions of facts. (e) raises questions about ontology and logic—the connection, among other things, of terms like 'not' and 'and' with "what is." Yet it is independent of those questions in one way. For the questions about ontology and logic would involve the use of 'and' in a sentence like 'Socrates is white and Plato is square' as well as in the sentence 'Socrates is white and Socrates is square.' But only the latter kind of sentence concerns us here. We are not concerned with the possible reference of logical terms but with the use of conjunction to express several facts about one thing and hence to reflect, along with the use of the same proper name, what some would call the "unity in difference" of Big Socrates. But more of this later. We might also note that on this view, if we acknowledge (d) and (e), then in addition to the predicative tie between universals and particulars one may think of a relation of *contains* between facts on the one hand and universals and particulars on the other, as well as between complex facts and simple facts. Again, I do not think it is crucial here whether such relationships can be expressed in one's language or show themselves or what have you. The point is that, taken as entities, facts stand in these relations to "things," things they contain as well as things they are contained by.

On the second view, Big Socrates is considered to be a composite of universals. By "composite" one does not mean a class or collection but a number of universals connected by some structural tie or relation, ontological glue as it were. Thus, to borrow Bradley's phrase, things are "qualities in relation." But this relation, like predication on the first alternative, is not a relation among relations. To think it so, on either alternative, is to invite a puzzle associated with the name of Bradley. This is the point of calling such relations "structural" or "ontological ties." The notion is that such relations are not themselves further "entities" which the composites contain and thus, in turn, require to be connected with the other constituents. One may offer the second alternative for two reasons. First, on it the analysis of Big Socrates involves only one kind of entity and a structural relation; whereas the first alternative requires the same kind of entity, a structural relation, and a further kind of entity as well, bare particulars. Second, one may feel that the additional entities of the first view, bare particulars, are neither comprehensible nor experiencable, while universals are both. The defender of the first view claims this second contention is mistaken. Thus one crucial issue involved centers around the analysis of what is meant by holding that bare particulars and, alternatively, universals, are objects of acquaintance. This issue we cannot take up here; rather, we shall take up further arguments seeking to establish the *existence* of bare particulars as necessary to account for what we are presented with (Big Socrates and Big Plato), independently of the question of whether or not we are acquainted with the elements to which analysis leads. While what follows may not exactly reproduce the language of any specific formulations of the arguments, it will, I believe, do justice to them as well as refute them.[1]

To get at some of the various points involved we shall, in this paper, omit any consideration of relations and consider the only nonrelational properties that Big Socrates and Big Plato have to be whiteness and squareness. They are then both combinations of these two universals. As such they do not differ with respect to their constituents. Or, as Moore would have put it, they do not differ conceptually. The protagonist of bare particulars may then ask what accounts for their difference. In doing so he thinks the question rhetorical for, as they do not differ in *anything*, they cannot differ. On his analysis the two patches would differ in that they contained (or were reduced to) different bare particulars. The bare particulars are the *things* wherein the difference lies. But cannot one reply by observing that the bare particulars differ, not in anything, but "simply differ"? Why not hold that Big Socrates and Big Plato are "simply" two different combinations of the same qualities? They differ even though they do not contain different constituents, do not differ in any respect. The bare particular analyst purports to uphold numerical difference in contrast to conceptual difference. Yet he seems to insist that for Big Socrates and Big Plato to differ they must differ in a respect, albeit not differ in a concept (universal) or conceptually. He turns their numerical difference into an entity. One wonders then if the bare particular analyst, in spite of what he says, has fully separated numerical from conceptual difference.

[1] Such arguments, as they appear in Moore and Russell, do not argue for "bare" particulars but for "particulars" and "numerical difference." But they also serve as arguments for bare particulars. One will find these arguments in B. Russell, "On the Relations of Universals and Particulars," *Logic and Knowledge*, ed. R. C. Marsh, London, 1956; G. E. Moore, "Identity," *Proceedings of the Aristotelian Society*, Vol. 1, 1901. Later versions of these arguments and the acceptance of the first alternative occur in papers by G. Bergmann in *Meaning and Existence*, Madison, Wisconsin, 1959; and *Logic and Reality*, Madison, Wisconsin, 1964; and by E. Allaire, R. Grossmann, and H. Hochberg in *Essays in Ontology*, The Hague, 1963.

For we might feel that to "differ conceptually" has a broader sense, to differ in a respect or constituent, and a narrow sense, to differ in a universal or quality. As opposed to this, to differ numerically is not to differ in anything but just to differ. This may be overlooked since the bare particular analyst, in one sense, identifies Big Socrates with Socrates and, in another sense, identifies the former with the facts about the latter. Be that as it may, for the bare particular analyst the two bare particulars, Socrates and Plato, are simply different without being different in any thing or respect. But why deny such simple difference to Big Socrates and Big Plato? To do so is to hold that only simples can differ "simply" or numerically. One hesistates to suggest that the term 'simple' provides a clue in that numerical difference is thought to be simple difference and, hence, only simples may differ simply or numerically. Perhaps the suggestion will not seem too outrageous when it is taken together with the idea that a complex of properties is either a class of such properties or something like a heap of objects. Russell, after all, did speak along the former lines,[2] or, at least, can be taken that way. If a bare particular analyst thinks of the second alternative in this way, then it would explain the belief that one "complex" of properties must differ from another in a constituent. But complexes, one might insist, are not to be thought of in that way. Just as two bare particulars may exemplify, by courtesy of predication and universals, one and the same property, there may be, by courtesy of universals and the combinatorial relation, two or more complexes of the same properties.

At this point another argument may be raised. The proponent of bare particulars names only simples. The proponent of property complexes names such complexes. Holding that only simples may be named, the former rejects the latter's view. But, why hold that only simples may be named? The proponent of bare particulars might hold to a "picturing" principle of language. Under this principle he would hold that the primitive simple terms of his "improved language" provide a list of existents since the simple terms correspond to simple objects, and ontology is a search for such kinds of simples. Involved in this are two ideas. First, one has reached simples when one has "analyzed" an entity down to constituents that need and can only be named but not described. Second, when one has shown how to connect a sign or sign complex to facts or things by means of other signs, the sign in question is not a name of a simple.

One says "need only be named" to convey that when one is acquainted with an object one can label it and need not describe it. One says "can only be named" to insist that to know what white is, in the sense of 'know' relevant to these issues, is to be acquainted

[2] B. Russell, *An Inquiry into Meaning and Truth*, Baltimore, 1962, p. 93.

with the referent of the term 'white.' Hence, while one may hold that one can know that something has the complex property of being a white-square if one knows that it is white and that it is square, any attempt to say something similar about white will involve, implicitly or explicitly, the use of empirical correlations between white and other qualities (of the same or of other entities) that are said to "define" it. No such empirical correlations are involved in "reducing" white-square to white and square.[3] Nor would a definite description serve as an analysis or definition. For one could be acquainted with (know) all the properties employed in a definite description without being acquainted with (knowing) what fulfills it. Thus, for someone who is not acquainted with whiteness, one can say that he could know what it is to be white in the sense of knowing a unique description of it without knowing what it is in another sense. A definite description may be thought of as a sort of recipe for arriving at knowledge by acquaintance, where there are things to be acquainted with. Not being "definable" or "analyzable," whiteness can only be named, if by the use of a sign that refers to an entity one is to "show" that one is acquainted with the referent and, hence, knows what it is. The same is thought to hold for bare particulars. To give the qualities that relate to a bare particular by predication is to say what it is in one sense. But those who advocate such things seem to feel that, in another sense, one cannot say about a bare particular what it is, just as one cannot say about whiteness what it is. One only knows it by being acquainted with it, and this is reflected by naming it. This shows the difference between Big Socrates and Socrates, for in the case of the former one says what he is, not just by listing his qualities but by bringing in Socrates and the tie of predication. Big Socrates is what he is in virtue of something else, Socrates. Hence, he can be dismissed. But about Socrates, in turn, this procedure will not do. He cannot be analyzed and to know him is therefore not to know his constituents but to be acquainted with him. In order to show this, to repeat, one must name him, just as to name him one must be acquainted with him. All this explains, in part, why some speak of both predicates and proper names as names or mere labels, of acquaintance as opposed to description, of meaning as reference, of existents as simples. This is why Big Socrates becomes Socrates and the qualities of the former, perhaps naturally thought of as parts in some sense, become things related to a bare particular to form still a third kind of thing, a fact. In this pattern we also see why it becomes important to speak of being acquainted with bare particulars on the analogy of being acquainted with properties like whiteness. That this is confused,

[3] Here questions about the nature of "analysis" could be raised. But, as I indicated, these are being side-stepped in this paper.

misleading, and false I cannot argue here. I simply note that it is integral to the bare particular analysis.

As opposed to a predicate like 'white' and a proper name like 'Socrates' both a predicate like 'white-square' and a sentence like 'Socrates is white' would be connected—one to a fact, the other to a complex property—by means of other signs. Thus just as sentences can be true or false, complex predicates may have been exemplified or not, but whiteness must have been exemplified in experience for one to know what the term 'white' connects with. Yet only in the case of the complex predicate would we speak of definition. One may speak of a sentence as referring to a fact in virtue of its signs referring to constituents of that fact, but these signs do not define the sentence. In both cases, however, one speaks of constituents as providing the elements of analysis and of meaning. This suggests two things. First, rightly, that to define, to analyze, and to give meaning are three notions and not one. Second, misleadingly, since Big Socrates, on the second view, is a complex of properties, one must think that the term which refers to him does so via the terms referring to his constituents, as in the case of sentences and complex predicates. A bare particular analyst might think along this latter line since, on his own analysis, Big Socrates may be referred to as a sort of complex fact. He might then think that on the second alternative a white patch is also a fact, as it is a complex of universals in a structural relation. Being a fact it should not be named, for facts are not named but are referred to by sentences via the referents of terms in the sentences. One might even argue that the second alternative obscures the factual nature of Big Socrates, whereas the bare particular analysis clearly distinguishes between the bare particular, the qualities, and the facts to which they give rise. What, one may ask, distinguishes Big Socrates from the fact(s) that he is white and square if he is a composite of white and square? He cannot be a constituent of such a fact in the sense in which a bare particular is. But does this say any more than that Big Socrates is not a bare particular? One cannot expect that on the second alternative there will be facts, as entities distinct from bare particulars. On the bare particular analysis a third entity, a fact, arises from the predicative connection of two simples. On the alternative analysis, since the subject is considered as a complex of qualities, no third thing comes about. The fact is, as it were, compressed into Big Socrates. But this is just a metaphor. What it reflects is a concern with the treatment of the sentence 'Big Socrates is white' on the second alternative. The bare particular analyst believes that this sentence must become analytic or tautological on the alternative analysis. This is a crucial reason why bare particulars are introduced. For if to know what a thing is, is to know its constituent properties, then that a thing exemplifies a property follows from its being "what it is." Thus to predicate of a subject a part of it is to be redundant and statements of fact become echoes of the pronouncement that "a thing is what is." This belief is mistaken. It arises in part, as did an earlier misunderstanding we considered, from thinking of a complex of qualities as a class of qualities and of predication as reduced to class membership. If one held that Big Socrates was a class of qualities, the class whose members are white and square, and that to say that he is white is to say that white is a member of that class, then there would be ground for the fear that the sentence 'Socrates is white' has become a tautology or analytic truth. This would also raise the question about the difference of Big Socrates and Big Plato. Furthermore, in its way it offers a definition of the term 'Big Socrates.' But this is not the view that we are considering. Our second alternative does not identify Big Socrates with a class of properties but with the members of such a class in a unique kind of relation. Yet, on it, one might still think that since Big Socrates was analyzed into a set of qualities in a unique relation, the term 'Big Socrates' *meant* white and square in a relation. One is then tempted to hold that to say 'Big Socrates is white' is to say that a combination of white and square contains white, and this is a bit redundant. It seems analogous to saying, on the bare particular view, that the fact composed of white and Socrates contains white rather than saying that Socrates is related to whiteness by predication. One also notes that to consider the term in this way neglects the difference of Big Socrates from Big Plato, since this difference is not considered as a constituent element. Thus we are led back to the mysterious bare particular which provides a constituent element to account for numerical difference. But this is an illusion. We have already seen in the case of facts that analysis is one thing and definition is another, although this is obscured by speaking of the meaning or reference of the sentence in terms of the meaning or reference of its constituent signs. The point is that to analyze our white patch is not to define the term that refers to it, nor is it to say what the term "means" in the sense of uniquely specifying its referent. One may think that one must define the term 'Big Socrates' or uniquely specify its referent if one analyzes the white patch referred to since one may hold (a) that to analyze a thing is to say what it is and that is to say what makes it different from anything else, and one may wonder (b) how otherwise could one attach the name to the referent. But this just repeats, first, the assertion that only in the cases of simples can something just be different from something else and, second, the "picturing" principle. Similarly, one would do nothing more than repeat the picturing principle by objecting that to refer to the complex thing, Big

Socrates, by a simple term is to acknowledge that complexes exist in the sense in which simples do, since both are referred to by simple terms. Even if one names a complex and refers to constituents of it by simple predicates one might still point out that one type of term refers to a complex entity and one to a simple one. Language simply does not picture this distinction via its simple and complex terms, although it may be thought to do so in another way, since all proper names would refer to complex entities. In any case the picture principle, being at issue, resolves nothing.

Be that as it may, let us return to the accusation that on the second alternative a sentence stating that Big Socrates is white becomes analytic or tautologous. Actually, a similar concern may be voiced about the bare particular analysis. Some may feel that if 'white' refers to a property of a bare particular and a proper name names the bare particular exemplifying the property then to use the two signs to say that the particular is white is to utter a tautology. I am speaking here without considering at all the complications that arise from the use of the predicate term on other occasions. It is clear that the form of the sentence 'Ws' is not analytic. It is also clear that one may say, on a bare particular analysis, that it is possible for the bare particular in the fact of which we speak not to be there, not to exemplify white. What strikes one as peculiar, nevertheless, is twofold. First, one suspects that the bare particular was introduced simply to carry, as it were, the property, and hence one is not told anything when one is told that it exemplifies the property. Second, one notes that given that the proper name names a constituent of a fact and a predicate refers to a quality of the same fact, then it follows that the particular exemplifies the property. The first point merely raises a suspicion about bare particulars. The second points out that what is tautological is not the ascription of a quality to a particular, but a conditional statement that would assert that the ascription holds if certain terms are tied to certain things. A similar point may be made about the second alternative, for, given that Big Socrates has been analyzed into a set of qualities, it would seem redundant to specify a member of the set. In a way this would amount to seeing that 'Big Socrates is white' follows from 'Big Socrates is white and square' for the analysis is expressed in the true statements ascribing qualities to him. The point, again, is that the predicates do not *give* a meaning or referent to the proper name, for the latter *refers* to, and in this sense only means, the complex thing we see. That the complex thing is composed of certain qualities does not imply that each quality forms part of the meaning of the name. A sentence is analytic due to the "meaning" of its terms when one term enters into the definition of another or occurs as part of a complex sign like 'All bachelors are unmarried,' where 'bachelor' is defined as unmarried man, or 'All red apples are red,' where the sign 'red' is part of the sign complex 'red apple.'[4] What is involved is the specification of the meaning of a sign, even as reference, in terms of the meaning or reference of other signs or simply being an instance of an analytic form. This does not occur when complexes are named on the second alternative, although one may be easily led into thinking it does, or should, due to (a) holding that meaning is reference and (b) accepting the picture principle. Analyticity, we must recall, is a matter of language.[5]

That a complex thing like Big Socrates is composed of some qualities rather than others is expressed in the true sentences about it. But one must not think that because one uses a proper name for the thing, the name must refer to something other than the composite of qualities. Proper names are expressions that can be used to refer to things without specifying properties of them. It is interesting to recall that, except for certain trivial cases, identity statements holding between a proper name and a Russellian description are synthetic. This feature of the relation of descriptions to names reflects the same fact as the use of a name for a composite of qualities without turning statements about the composite into tautologies. It even leads one to wonder if a nominalist like Quine, when he proposed the replacement of proper names by definite descriptions involving individuating properties like Pegasizing, was not ultimately worried about bare particulars, for such a gambit may be considered as a feeble version of the second alternative. Perhaps, then, bare particulars simply result from a hypostatization of this feature of proper names. And that is not surprising in a decade when one gets at ontology through language, improved or otherwise.

The proponent of bare particulars might find that the second alternative simply amounts to a way of having the benefits of bare particulars without the odium of acknowledging them. Hence, he might feel that the second alternative does not *really* differ from his analysis. But what is it, in such matters, to say something different? Supposing that both views could be upheld, would this constitute saying the same thing? This is perhaps what the objector has in mind. One alternative seems to be recognizing that there is a certain kind of thing, a bare particular, that the other denies. But then, what is it to be a bare particular? The merry-go-round would begin if the proponent of bare particulars held that Big Socrates, on the alternative view, is really a bare particular since the

---

[4] This is not put precisely, but the point is, I believe, clear.
[5] Russell has argued that one can use proper names to name complexes of qualities without turning the relevant assertions into analytic statements, *ibid* p. 122. Though part of my argument is similar to his, I cannot discuss here how they differ or why I take his view to be confused.

name 'Big Socrates' refers to one instance of a combination of qualities while the name 'Big Plato' refers to another. And, after all, bare particulars are just that. But then, why talk about bare particulars at all if they are just combinations of qualities? In any case, there are other differences.

First, a bare particular need exemplify only one universal. This is one thing involved in the doctrine of logical atomism. On the alternative view, since a thing is a combination of qualities, there could not be a thing with only one quality. Thus there is a difference in the notion or explication of "thing." Furthermore, the proponent of the second view might point out that to speak of a thing with only one quality is to come close to identifying the bare particular analysis with yet another view, which we shall soon consider, which holds that the elements of analysis are neither bare particulars nor universals, but instances of universals.

Second, predication, on the view that a thing is a composite of qualities, is one thing as it holds between a composite and one of its constituents and another between a simple first level quality and a simple second level quality, if there are any of the latter. On the bare particular analysis the *same* relation *seems* to hold between particulars and first level qualities and between those and second level qualities. Whether there is only one such "tie" on the bare particular analysis is a question; but if so, this would constitute a further difference, provided there are simple second level universals.

Third, a bare particular, *qua* bare particular, may exemplify a succession of different qualities, i.e., it may persist through change. That such a bare particular could not be "recognized" or that there may be other insurmountable difficulties in saying this is beside the point. A composite of qualities, as such, could not persist through a change of quality. It would make no sense or, perhaps better, I take this to be one of the ideas involved in the use of 'composite.' Qualitative change would result in a new combination, just as on the bare particular view it would result in a new fact. All that can be said on the combinatorial view is that the thing before and after the change shares some, but not all, qualities. Something else can be said on the bare particular view, i.e., something in addition to any quality is one and the same. Bare particulars, as we might suspect, are adolescent substances, even for one who starts out by talking about phenomena. The alternative view is thus in closer contact with one classical motif of those who reject realism: things do not persist through change. Perhaps this feature of the combinatorial view supports the notion that upon such a view statements about Big Socrates being white and square are analytic. For, just as a combination of qualities cannot, so to speak, change a constituent, it would not be the same combination if "it" had a quality other than those it has. Thus one may be led to think that to say of such a thing that it has a certain quality is to state a necessary or analytic truth. But, again, this all goes back to not taking the meaning of the name to be given, in any sense, by the relevant predicates separately or in combination. Perhaps one ought to add the two trite observations that a combination of black and round could have been when and where Big Socrates was and that the term 'Big Socrates' could have named another composite. A proper name of complex entities, or of bare particulars, is simply an indicator like 'this.' One is almost tempted to say it has a referent but no meaning, for to talk of meaning and reference lies at the core of the puzzle. A proper name refers to a complex object. A list of true sentences ascribing properties to the object provides the elements in the analysis of the object. Thus they provide an answer to the question "What is it?" But to say what a thing is, if one uncritically thinks of meaning as identified with reference, is to say what the term which refers to the thing means. Recall the case of a fact and of a complex property where both mean and refer in virtue of their constituents. The bare particular analyst, having his special entity, takes it as the referent of the name and mentions qualities when asked "what the thing is." Hence, he keeps what is named separate from what is predicated quite easily. Not having such an entity, the proponent of the second alternative may succumb to holding that white is part of the meaning of the term 'Big Socrates' since what Big Socrates is includes white. Thus, as one proceeds in the analysis of Big Socrates, one uncovers more and more of the meaning of the term which refers to him. When one has gotten all his qualities any statement about him that ascribes a quality to him will be analytic. The meaning of the name changes as our knowledge progresses, while the referent stays constant. Thus the meaning "grows" until it "equals" the referent. Add relations in the form of relational properties and one is well on the road to Bradley's Absolute. The bare particular then emerges in a new and heroic role as a last line of defense against holistic idealism. Professor Bergmann, a contemporary exponent of bare particulars, is, interestingly enough, a logical atomist and, lately, a realist. But standing at opposite extremes both the absolute idealist and the proponent of bare particulars make an extremist mistake. To the question "What is Big Socrates?" we may answer either "this" or list his properties. One extreme, that of absolute idealism, is then to collapse the first into the second. The proponent of bare particulars, to preserve the integrity of the first reply, invents a special thing for the indicating sign to name. In doing what they do both seem to accept the illusion that if a term can have a simple indicating function there must be a simple indicated object about which we cannot say what it is, i.e., what it is composed of. The idealist starts with the indicating function of proper names but, unable to comprehend a pure this or bare substratum, he collapses this referring function into the specification of the properties of what is named. As it were, the "that" disappears into the "what." The proponent of bare particulars divorces, not the "that" from the "what," but the function of the term 'that' and endows it with existence. To make such a thing palatable and comprehensible, since he cannot say what it is, he convinces himself that he is acquainted with it. The lesson is that if we separate the purely indicating function of proper names of complexes from the question of specifying the composition of such objects, one can reject bare particulars as well as holism. It is, I believe, that simple.

If, after listing all the properties of Big Socrates, one "thinks" of him in terms of those properties—if the term 'Big Socrates' no longer functions as a simple indicator of an object—then, perhaps, one's statements about him become analytic. But put this way, what is the problem? Is the sentence 'Big Socrates is white' analytic? One may reply that it all depends on what you *mean* by the term 'Big Socrates.' The crucial point is that *this is not to say* that it all depends on what the term 'Big Socrates' refers to. To think otherwise would indeed lead one into holding that to ascribe a quality to an object is just to say, in different words, that the thing is what it is. Nor should the matter be further confused by the fact that we are led to indicate an object we are acquainted with by a name only when apprehending some qualities of it. We do, after all, notice something about what we notice. But this does not mean that the qualities we notice provide meaning for the proper name used as an indicator of the object. After all, something of the same sort is acknowledged by the advocate of bare particulars when he claims that he perceives them never, as it were, by themselves, but always attached to qualities.

The advocate of bare particulars may still complain that, while predication is clearly a relation between two entities on his view, it is not at all clear what it is on the alternative view. Two comments. First, what is involved, once again, is the picture principle. On the bare particular analysis each term of the sentence 'Socrates is white' indicates something in the fact. On the alternative view the copula does not reflect an ontological tie between a bare particular and a universal. Since the proper name names complexes, the tie is already partially reflected in the use of the name. The copula serves in the sentence to enable one to specify what qualities are contained in the complex and hence also reflects the combinatorial tie. Second, in spite of explicitly recognizing that an ontological tie, like predication, cannot be thought of as a relation among relations, the bare particular analyst may, in a way, do just that. One argument for bare particulars involves holding that a relation cannot distinguish objects since these objects are needed to stand in the relation. The idea is that the relation requires terms. Thinking of predication in the same way, one might be led to introduce bare particulars along with universals as distinct entities required by the relation of predication. The combinatorial view would not do, since it might be thought that on it we do not have two distinct things to stand in a relation. On the one hand, this takes us back to the problem about analyticity; on the other, it presupposes that predication is like, say, "left of."

The third alternative analysis actually comprises three possible views. On one, Big Socrates is analyzed into instances of qualities and its proponent recognizes both instances and qualities themselves; on a second, only instances are acknowledged; on a third, the instances are themselves analyzed into composites of qualities.[6] I wish to consider only the first variant. According to it, Big Socrates is a composite of an instance of whiteness and an instance of squareness. Big Plato, in turn, is composed of numerically different instances of the same qualities. On such a view, then, we have in addition to the complex entity Big Socrates and the simple entities, qualities and instances of qualities, a combinatorial tie that unites the instances, and a predicative tie between an instance and the quality of which it is an instance. To a bare particular analyst such a view has several difficulties.

First, it requires two ontological ties. Second, this view, like the second alternative, ends up by turning the sentence 'Big Socrates is white' into an analytic truth. Third, as long as such a view acknowledges qualities, its instances might just as well be bare particulars. The use of instances is significant only when used as a nominalistic gambit. Fourth, the instances are themselves disguised combinations of a bare particular and a universal and hence are complex, not simple. Two instances of white, like two white patches, have something in common and yet differ numerically.

In reply, a proponent of the third view might argue that if Big Socrates, on the bare particular analysis, is resolved into a conjunction of facts, then there is, as we noted earlier, a combinatorial function played by conjunction and the use of a single subject to "exemplify" all the properties of the object. This would be especially significant if, due to the picture principle, the bare particular analyst was led to minimize the distinction between logical and descriptive signs by, as some put it, "ontologizing logic." To

---

[6] The third variant may be attributed to Moore at a certain stage; see my "Moore's Ontology and Non-Natural Properties," reprinted in *Essays in Ontology*, *op. cit.*

"ontologize logic" is to have the primitive logical signs and features of a language refer to or reflect some thing or feature of the world. In any case, the combination of a bare particular and a predication relation that can hold between one given subject and many qualities plays a role analogous to the combinatorial relation on the third view. On the third alternative, we most note, any given instance of a quality is only related by predication to that quality and to no other. Hence, the connection of particular to quality and of particular to particular must be achieved by two relations. The merit of the second alternative is to dissolve the one into the other by avoiding instances. On the third alternative, as on the second, facts, simple or complex, do not lend themselves to becoming entities; for the relation between a whole and a part, Big Socrates and a quality or instance of which he is composed, is not a connection between two simple but different kinds of things which gives rise to a third distinct and complex thing, a fact. One is just uncovering, so to speak, what is there. This is what gives rise to the fear of analytic statements. But this we have discussed at length. However, perhaps three more observations may be tolerated on that topic. First, to hold on either the second or third alternative that there is something white is, I believe, obviously not analytic. Yet that would follow from the statement that Big Socrates is white, where the term 'Big Socrates' is a proper name. Hence, to consider the former analytic is to hold that a synthetic statement logically follows from an analytic one. This is peculiar. I find the peculiarity to rest in thinking that the sentence 'Big Socrates is white' is analytic; others, of course, may find the queerness in the second and third alternatives. Second, it is quite understandable that the bare particular analyst finds the sentence 'Big Socrates is white' analytic on the alternative views; for recall that his sentence is 'Socrates is white' and, as we saw earlier, the only other reflection of Big Socrates would be in the conjunction 'Socrates is white and Socrates is square.' From this it logically follows that Socrates is white. Hence, he might naturally think that to ascribe whiteness to something after declaring it to be a complex of qualities or of instances is to do so analytically. Third, on the last alternative the question of analyticity is further complicated by the connection of an instance to its quality, for a sentence stating this connection might be thought to be analytic.

A consideration of the third alternative has thus pointed up the combinatorial role of bare particulars. Thus, they are more than hypostatizations of numerical difference and of the indicating function of proper names. A further aspect of this is revealed if we consider, as a possible variant of the bare particular analysis, the notion that each quality must have its own bare particular. On this view a bare particular analysis must explicitly have recourse to a combinatorial relation to get at the "fact" that Big Socrates is both white and square. Here, indeed, instances and bare particulars would meet, for the key difference is that instances can exemplify only one property. (There would still be the question of a bare particular changing quality, which would, I take it, not fit the "logic" of instances.) The question of deciding between two such alternatives may not be relevant to the existence of bare particulars. It is relevant to the claim that they are experienced or directly apprehended. But that claim involves a further set of issues.[7]

Terms like 'nominalist' and 'realist' are freely used. In some circles to be a nominalist is like being a "revisionist" or "deviationist" in others; among other philosophers a realist or Platonist is one who requires analysis of a nonphilosophical kind or, at least, a "shave." In spite of this I cannot refrain from noting that of our three alternatives there are some senses in which the bare particular analysis may be held to be the one closest to nominalism. I mention two. First, qualities require a substratum. On the second view they may be said to compose it, and on the third the qualities are distinct from the instances of them. Second, remove the qualities, as distinct from the instances, from the third alternative and we have, in substance, the bare particular analysis. Perhaps medieval historians have a word for it when they speak of "moderate realism."

[7] These issues are discussed in my "On Being and Being Presented," to appear in *Philosophy of Science*, and "Ontology and Acquaintance" to appear in *Philosophical Studies*.

COMMENTS

RICHARD SEVERENS

It is recommended that we construe objects such as sense data as composites of qualities, rejecting thereby both the view that objects are bare particulars exemplifying qualities and the view that objects are composites of instances. Among the several questions which naturally arise, perhaps the most interesting is whether the operation of predication can be comfortably accommodated by any of the three views. This is a question of some import because predication (or its equivalent) seems to be the logically primitive operation for formulating discourse about objects. That is, discourse about objects consists either in predications or devices which presuppose predications. Moreover, predication seems to be necessary for the articulation of truth and falsity in discourse about objects. That is, the elements of discourse which are susceptible to truth or falsity are predications or devices which presuppose predications. In this way, truth-functional discourse about objects may be said to presuppose predication. Let us see, then, how predication fares with respect to each of the three views.

Immediately a distinction is needed. Predication is sometimes construed as the connection, whatever it may be, between an object and its qualities taken severally, perhaps as the relation of exemplification, perhaps as a nonrelational connection. Sometimes, on the other hand, predication is construed as the connection between an object and appropriate linguistic predicates, expressed by affixing such predicates to linguistic subjects. (It is Johnny, not 'Johnny,' of which 'blond' is predicated.) The latter conception of predication differs from the former in that one of its terms must always be a linguistic predicate. It is this latter conception of predication which is involved in what follows. (Sometimes, the results of affixing linguistic predicates to linguistic subjects are viewed as predications. For present purposes, such results will be regarded instead as expressions of predications.)

The view that objects are to be construed as bare particulars exemplifying qualities is not in any natural way receptive to predication. When we predicate 'blond' of Johnny, we do not predicate it of Johnny's bare particular, but rather of Johnny himself. And this is true of any relevant predication of quality words that we could make with respect to Johnny. We do not predicate 'blond' of something which has no color, 'tall' of something which has no height, 'heavy' of something which has no weight, 'roly-poly' of something which has no shape, and so on. The predication of any quality word with respect to Johnny is a predication which has as its subject, not Johnny's bare particular, but Johnny himself. That is to say, there can be no predications of quality words of which Johnny's bare particular is the subject. (If one thinks he detects echoes of Berkeley here, he is absolutely right.) Yet the most natural form of the view that objects are bare particulars exemplifying qualities turns in part on the analogy of bare particulars to linguistic subjects and qualities to linguistic predicates.

In order to avoid this, a *connection* may be introduced between Johnny's bare particular and his several qualities, which is an entity in its own right. It is most natural to view this connection as the relation of exemplification, whence expressions of predications will have the form: 'Johnny's bare particular exemplifies blondness.' But this leads directly to Bradley's regress. Instead, then, the connection might be viewed as a nonrelational connection. But this, so far, is to baptize, rather than to resolve, the issue.

There thus seems to be considerable difficulty in articulating predication for the first view, pending specification of some connection between bare particulars and qualities.

For different reasons, the view that objects are to be construed as composites of qualities is alien to expressions of predication. In order to avoid Russell's and other paradoxes, it is deemed necessary to type or stratify all expressions of predication. This takes the form of requiring that in every admissible expression of predication, the predicate shall be one type-level or stratum higher than the subject to which it is affixed. But no such typing or stratification can occur in discourse about objects, according to the view that objects are composites of qualities. Two possibilities arise. Either the manner of composition of the qualities—a class or relation or whatever it is that makes a given quality a component of a given object—is an entity in its own right or not. If it is not an entity in its own right, and objects are simply congeries of qualities, then every relevant predication will be such that the subject and predicate are at exactly the same type-level. Since Johnny is merely a congeries of qualities, every appropriate quality word affixed to 'Johnny' will be at the same type-level as 'Johnny.' Such expressions will be type-violating, and therefore inadmissible.

On the other hand, if the manner of composition of the qualities is an entity in its own right, something which collects the qualities, then the thusly compounded objects will be of one type-level higher than anything which is predicated of them. 'Johnny' will turn out to stand for a compound which includes or contains Johnny's qualities. Quality words affixed to 'Johnny' will thus be one type-level lower than 'Johnny.' Once again such expressions of predications violate the theory of types and are inadmissible. There can therefore be no admissible expressions of predications of quality words in discourse about such objects.

The view that objects are composites of instances is similarly hostile to expressions of predications. If such objects are simply congeries of instances, then all applicable descriptive predicates will be of exactly the same type-level as that to which they are affixed. If, on the other hand, the manner of composition of the instances is an entity in its own right, something in virtue of which the composite is compounded, then the compound will be one type-level higher than anything predicated of it. 'Johnny' will be one type-level higher than 'blond' or any other appropriate quality word. All such expressions of predications are type-violating and, thus, inadmissible. It turns out once again to be impossible to formulate admissible qualitative descriptions of such objects.

This situation could perhaps be remedied by replacing predications of quality words with predications of relation words. Thus, 'Johnny is blond' gives way to 'Johnny includes blondness' or 'Johnny contains blondness,' where the relation words 'includes' and 'contains' are predicated, not the quality word 'blond.'

But, beyond distorting the form of discourse descriptive of objects, this suggestion remains merely a suggestion until the appropriate senses of 'includes' or 'contains' are specified. And one may well entertain considerable skepticism as to whether they can be satisfactorily specified. The little word 'is' seems to be distinctly superior to these latter connectives. Moreover, it usually turns up in the specification of more exotic copulas, suggesting thereby that it is the primitive connective. Lastly, the predication of quality words seems to be logically more basic than the formulations entertained a paragraph ago.

Perhaps what all of this indicates is that any analysis of objects must square with the form of discourse about such objects. Perhaps, instead, linguistic innovation is indicated. It remains to be seen which is the case.

COMMENTS

J. M. SHORTER

Mr. Hochberg considers three accounts of what a white square is. It seems to me possible that, in so far as these accounts can be made intelligible, they must all be true. At any rate when Hochberg touches on this topic, what he says appears to be unsatisfactory. I wish therefore to argue that these theories must either be unintelligible or else compatible with each other.

In Hochberg's favored theory the expression that gives the trouble is the phrase 'combination of qualities.' It is, of course, not obvious what it means in the context of the theory. If we take as our starting point certain more familiar uses of the word 'combination,' then it is easy to arrive at a sense for the phrase 'combination of qualities' which is clearly not the one required. Consider the following sentences:

> Brine is a combination of salt and water.
> This screwdriver is a combination of a plastic handle with a metal shaft.
> Indignation is a combination of anger and a feeling that injustice has been done.
> A drive in golf is a combination of ninety-nine distinct movements.

In these cases the combination is of the same general sort as the ingredients. Brine, like salt and water, is a stuff rather than a bit of stuff. A screwdriver, like a plastic handle and a metal shaft, is a material object. A drive, at golf, is itself a movement. Instead of saying "Brine is a combination of salt and water" one could have said "Brine is a complex stuff, consisting of salt and water." One could say that indignation is a complex psychological state, a drive, a complex movement, and so on. Pursuing this line of thought one arrives at the conclusion that a combination of universals should itself be, if anything, a universal. In one sense of 'combination' of the qualities white and square, such a combination would be the complex quality, being white and square. But this sense will not do for the theory we are now considering. For, according to such a theory, there can be two or more combinations of exactly the same universals. But it is surely hard to make sense of the idea of *two* complex universals which are both identical with *the* complex universal, being white and square. Indeed it is probable that this sort of consideration has been one of the things at the back of the belief that the combination of universals theory is unacceptable. How, by combining universals, can one get something, like a patch, which is nothing like a universal? One might as well say that a drive in golf is a combination of eighty-eight muscles and fifty-two bones as that a patch is a combination of one shape and one color.

An objection of this sort can be overcome only by saying that the word 'combination' is being used in a different sense or, if this way of putting it is preferred, that a different sort of combination is involved. One might say that a drive in golf is a combination of x muscles and y bones in the sense that when a drive takes place these are the only things that move. One might call such a combination a movement combination to distinguish it from other sorts of combinations of muscles and bone like a human body. Can something similar be done for the phrase 'combination of universals'? Of course it can. One could introduce the phrase 'combination of universals' by laying down some such rule as the following: a thing may be called a combination of the universals ϕ and ψ if it is ϕ and ψ. If one does it this way, however, it becomes a truism that if Big Socrates is a white square patch, then it is a combination of white and square; and it looks as though all one has done is to introduce a new terminology. Instead of saying "pass me the right hand one of those two white square patches" one could say "pass me the right hand one of those two combinations of whiteness and squareness," and so on. But it is hard to see how the meaning of 'combination of universals' can be explained without this unwelcome result.

It is perhaps unnecessary to deal with the bare particular theory in the same detail as I have tried to deal with the combination of universals theory. Here I suspect I am not really far removed from Mr. Hochberg. Of some accounts of it I would say that no clear sense has been attached to the expression 'bare particular.' Perhaps Hochberg is, in his own way, saying something similar when he calls bare particulars *peculiar* entities. Other accounts, however, make the theory intelligible but truistic and not readily distinguishable, except in terminology, from the combination theory. Such an account is the one Hochberg mentions when he says:

> The merry-go-round would begin if the proponent of bare particulars held that Big Socrates, on the alternative view, is really a bare particular since the name 'Big Socrates' refers to one instance of a combination of qualities while the name 'Big Plato' refers to another. And, after all, bare particulars are just that (pp. 91–92).

What does Hochberg say in reply to such an imagined proponent of bare particulars? He first comments, "Why talk about bare particulars at all if they are just combinations of qualities?" To this question the reply might be, "But they aren't just combinations, they are *instances* of combinations." Such a reply would be made by someone who is (reasonably) using the phrase 'combination of qualities' to mean 'complex quality.' To explain the meaning of 'instance of a combination of qualities' it would seem necessary for him to say something like this: "If anything has properties x, y, z, it may be described as 'an instance of the properties x, y, and z,'" If this is done, it becomes quite clear that the phrase 'instance of a combination of qualities' means the same as 'combination of qualities' in Hochberg's sense of that phrase. Suppose that we used the phrase 'objective combination' for this sense; then the point of adding the adjective would be the same as the point of adding the expression 'instance of.' There also seems to be no reason why one should not convey the same idea by using the words 'particular exemplifying' instead of 'instance of.' If Hochberg can say, "Why talk of bare particulars at all if they are just combinations of qualities?" his opponent can equally say, "Why talk of combinations merely if these are particulars exemplifying qualities, *instances* of combinations of qualities?"

However, Hochberg has some reasons also for holding that the two theories *are* indeed different and incompatible. First he says:

> A bare particular need exemplify only one universal. This is one thing involved in the doctrine of logical atomism. On the alternative view, since a thing is a combination of qualities, there could not be a thing with only one quality. Thus there is a difference in the notion or explanation of "thing" (p. 92).

I should like to make one comment here. This appears to tie the bare particular proponent down very closely to the doctrine of logical atomism. But does such a proponent *need* also to be a logical atomist? If he does not, then he could, if he wished, hold the view that a thing is a bare particular exemplifying at least *two* universals. It may here be said that on the bare particular theory this requires a somewhat arbitrary stipulation, whereas on the combination theory no stipulation at all is required. But this is not so. A combination of qualities is a localized combination, and so the qualities which make it up become localized in being combined. There remains the possibility that a single quality may become localized in one or more places without being combined with any other quality. If the combination theorist wants to rule this possibility out, he needs the stipulation that localization is impossible without combination. Such a stipulation is surely the same stipulation as that of the bare particular theorist, but expressed in a different terminology.

Professor Hochberg's second reason is this:

> It [a bare particular] may persist through change.... A composite of qualities, as such, could not persist through a change of quality, It would make no sense or, perhaps better, I take this to be one of the ideas involved in the use of 'composite' (p. 92).

I find this passage puzzling. It seems to run counter to the view that Big Socrates is a combination. Squares can undergo change, and usually do; white ones are liable to become dirty. Big Socrates is a white square. Therefore, Big Socrates can undergo qualitative change; therefore, it is not a combination.

A consideration of the nature of proper names and of demonstratives like 'this' and 'that' may serve to throw some light on the situation and show that there is really no difference here between the two theorists. Suppose that we are at a naming ceremony. A baby is held up and the namer points his finger in the direction of the baby, the surface of the baby, the yelling of the baby, the baby's movement of its limbs, the current baby stage, the fusion of all babies,[1] and the objective baby combination. At the same time he says, "I name this John." What has now acquired a name? Nothing has, unless there is some convention about how we are to use the name 'John' in the future as a result of the ceremony. Our use will be different according to what entity has been named. If, for example, we had named the objective baby combination, then it would be appropriate later to say "bring me John" when we wanted the baby only if the baby was a remarkably unchanging one. If we had named the baby fusion, then the order "bring me a discrete part of John" would be obeyed if any baby whatever arrived. If we had named the yelling of the baby, then the sentence "John was two feet long" would be some sort of absurdity. In short, to acquire sense as a proper name in a naming ceremony, a word has to acquire, in that ceremony, a particular logical grammar. This is what is involved in its becoming the name of one type of entity rather than of another.

The reasonable conclusion, therefore, is that when Hochberg introduced the phrase 'Big Socrates' as a proper name, he meant it to be the name, not of a white square, but of a white square objective combination. Similarly, if the bare particular theorist says that Big Socrates can undergo qualitative change, it is reasonable and charitable to suppose that he is using the proper name or label, 'Big Socrates,' as a name for a white square and not for a combination. If they both use the name for a different type of entity, it is hardly surprising that they seem to contradict each other, when in fact they both say something true.

[1] That single although discontinuous portion of the spatiotemporal world that consists of babies, cf. W. Quine, *Word and Object*, New York, 1960, p. 52.

Concluding Note

In his original paper Mr. Hochberg used the expressions 'white square' and 'white square patch.' I assumed, in my reply printed above, that he was using such expressions in their ordinary sense for publicly observable objects like a patch in a pair of trousers. I understand, however, that he intended in fact to be discussing sense-data. This does not, I think, seriously affect my comments. It is, however, worth indicating how they would need to be altered to apply to sense-data. I shall assume that the notion of a sense-datum is an intelligible one.

The expression 'white square sense-datum,' unlike 'white square patch,' is a technical one not found in ordinary language. It has not got a clear logical grammar and needs to be given one before its sense can be clearly fixed. Hochberg would, I think, agree that, in his sense of 'sense-datum,' I am having a sense-datum when I have an after-image. Suppose that I have a white square after-image and that this after-image becomes green. One might in such circumstances say that my original white after-image stage had ceased to be and had been replaced by a green after-image stage. It is surely clear that we might give 'sense-datum' such a sense that an after-image would be an example of a sense-datum; or we could give it a rather different sense in which only an after-image stage (with its temporal boundaries determined by qualitative change) would be a sense-datum. Following Russell,[2] Hochberg, it seems, wants to use 'sense-datum' in the latter sense. If so, it is a truism that a white square sense-datum is a combination, for Hochberg's sense of 'combination' also excludes the possibility of a combination changing. But this does not mean that a philosopher who denies that a sense-datum is a combination, in Hochberg's sense of 'combination,' is in disagreement with Hochberg. It is more likely that he is using 'sense-datum' in a sense in which an after-image, not an after-image stage, is a sense-datum. In this sense of 'sense-datum' it is obvious that a sense-datum can undergo change and is not a combination (in Hochberg's sense). For in this sense of 'sense-datum,' what Hochberg calls a sense-datum is a sense-datum stage.

In short, the only emendation I need to make to my original comments is to allow for the fact that we have a certain freedom in what sense we give to 'white square sense-datum,' since its sense is not determined by ordinary language.

[2] E.g., see the discussion at the end of the second lecture in "The Philosophy of Logical Atomism," *Logic and Knowledge*, London, 1956, p. 203.

REJOINDERS

HERBERT HOCHBERG

Mr. Severens poses several criticisms of views we have considered. First, he argues that when we predicate we do not predicate of a bare particular, Socrates, but of something else, Big Socrates. His argument consists of citing several examples. One might be tempted to say that what Severens says is irrelevant since what *is* at issue is the analysis of the sentence 'Big Socrates is white' in terms of an ontology including bare particulars—what *is not* at issue is that 'Big Socrates' is the "name" that occurs in the sentence we start with. Severens goes on to consider an analysis in terms of a bare particular, a universal, and a tie of predication. He rejects this view since it leads to Bradley's regress. He seems to feel that Bradley's regress is involved because (a) exemplification stands in some relation to a bare particular and a universal, and (b) the phrase 'Johnny's bare particular' indicates that there is a relation between Johnny and his bare particular that must be introduced. But it is question-begging to take the sentence Severens uses as a paradigm for dealing with predication. The advocate of bare particulars considers a sentence of the form '$B(j)$,' where 'B' names a quality and 'j' names a bare particular, to assert that j exemplifies B. This, indeed, raises questions about bare particulars and the relation of such things to things like Johnny, i.e., the connection between Socrates and Big Socrates; but in considering predication as a unique kind of "tie" it avoids Bradley's regress. Thus, while there is a question involved in (b), above, the regress of (a) is avoided. Severens insists that one cannot avoid the problems posed by Bradley's regress by holding that predication is a quite different sort of thing from ordinary relations or entities. For bare particulars there must be *some thing* which connects them to qualities. Part of his argument amounts to insisting that one predicates of Big Socrates, not of his bare particular. Predicating of Big Socrates does not seem to involve some *thing* to connect a quality to a subject, apparently, because we *predicate* qualities of Big Socrates whereas, since we do not predicate qualities of bare particulars, we must *connect* qualities and bare particulars by *some thing*. This, as I suggested above, is irrelevant; it confuses what we seek to analyze with a proposed analysis or, at least, rejects an analysis by too strict obedience to the ordinary idiom. The other part of his argument amounts to his insistence that differentiating predication from other relations and entities baptizes rather than solves an issue. But this is so only if one insists that the issue must be resolved by considering predication as a relation among relations and a substantive among substantives. This is, in effect, to do exactly what Bradley did instead of to draw any lessons from Bradley's puzzle. To say that this "merely" baptizes an issue is misleading. What is baptized is a further sort of "thing" that one feels forced to acknowledge in order to avoid certain problems. And, after all, this is what one does in metaphysics. To put it another way, just as the baby is crucial in a baptism ceremony, the "discovery" that exemplification must be fundamentally different from other relations is what is crucial here. Thus, one may be said to baptize by using the phrase "ontological tie," but what is baptized is something other than an issue. All this is easily seen when Severens says that predication fails "pending specification of a connection between bare particulars and qualities." What could he possibly mean by "specification" other than the citing of some ordinary relation?[1]

His next arguments are directed against the view that objects are composites of qualities. He sees a type violation in such a view. I do not. The only explanation I can think of for his criticism is that he thinks of composites of qualities either as qualities in the sense in which one might say that the quality red-square is a composite of the qualities red and square, or as classes of qualities. In the first case, the composites would be of the same type as the entities of which they are composed. In the second case, the particulars would be of the same type as qualities of the qualities of which they are composed. One *might* see problems here about types. But as neither of these views is involved in what I said, we need not go into such matters. Severens might feel that to speak of composites as I did requires a mysterious way of combining qualities into things. But, then, this is simply to re-raise his worry about what predication is for bare particulars by worrying about what combination is on the alternative view, for *combination* on this view plays the role of the ontological tie.

As Severens wishes to know what predication is, Shorter wants an explanation of combination. But as Shorter poses the question, he desires, unlike Severens, an explanation of what metaphysics is. Behind his question is the view that the three alternative ontologies, call them (α), (β), and (γ) all say the same thing since they all agree that we are talking about a white square patch. Hence, they do not disagree as to what is there but only as to how to speak about what we all know to be there. Metaphysical or ontological analyses thus

---

[1] In my paper and this discussion I have, unfortunately, used "predication" and "exemplification" interchangeably as well as using the former for both the linguistic connection between signs and for the behaviour involved in connecting signs.

reduce to proposals for esoteric ways of speaking. In its way, this argument incorporates a version of a principle of verification as a criterion of meaning. It is not put, explicitly, as with the early positivists, or confusedly lumped with the process of inquiry, as with the pragmatists, but is buried under countless examples of ordinary uses of terms and requests for explanations in terms of such uses. Thus, what is argued is that since (α), (β), and (γ) agree that there is a white square they do not disagree about anything that is a matter of fact. What can they then possibly disagree about, except words? This comes out quite clearly when Shorter suggests that the combination "theory" simply adopts the rule that some thing is a combination of qualities ϕ and ψ if it has ϕ and ψ. Similar reductions would be involved for the other "theories." His suggestion is not illiterate only if he implicitly accepts the line of thought just mentioned. Clearly, two ontologies cannot disagree about the facts from which they start. Their disagreement is about their "ontological analysis" of such facts. But if one forces philosophers to disagree about such ordinary facts in order to disagree at all, one implicitly accepts a criterion of meaning that rejects metaphysical questions. The proposal that to differ meaningfully is to differ about ordinary facts is just one way of putting a verification principle. The issue then reduces to explaining how ontological positions do differ; in brief, to get Mr. Shorter to "see" what a metaphysical question is. Here one runs the risk of accepting Shorter, or one who shares his outlook, as an embodiment of a meaning criterion. The metaphysician need not do that, for although he is obliged to try to explain what a metaphysical question is, he is not obliged to succeed.

One might begin by suggesting that (α), (β), and (γ) differ about what sorts of "entities" or "elements" the parts of a sentence which asserts that the patch is white and square refer to. In a perfectly ordinary and obvious sense such a sentence is *about* the patch. But the metaphysician wishes to inquire into the connection(s) between the sentence and the patch. This leads to talk of *facts* and of *analysing* the patch into elements like qualities, bare particulars, etc. Ontology is not just about words nor about facts and things as we are ordinarily concerned with them in our everyday activities. It is concerned with the relation of words and facts: with what sorts of things words or sentences must refer to in order to answer certain questions. This is what has led some to speak misleadingly of a reference theory of meaning, both pro and con. A concern with the reference of words and sentences is just another way of expressing a concern with metaphysical or ontological questions. It is thus no accident that so-called Oxford philosophy has directed a sustained verbal discharge against "meaning as reference" and in favor of explaining meaning by illustrating ordinary use.

We start with ordinary things, be they physical objects or phenomena, and ask about the truth of sentences that are about those things. Take two sentences about Big Socrates: 'Big Socrates is white' and 'Big Socrates is square.' One asks in virtue of what connection between these sentences and Big Socrates are they about him and true. One could simply say they are true because of Big Socrates. But this does not quite do since, first, it is not Big Socrates but Big Socrates being what he is—white and square. This leads one to speak of facts. Second, while in one sense Big Socrates is the thing that is involved in the truth of the two sentences, in another sense what makes the one true must be different from what makes the other true. Since they say different things, they indicate, in some sense, different aspects or things about the world. This, too, leads one to speak of facts as distinct from ordinary things like Big Socrates. Facts, as it were, constitute the first step in the metaphysician's concern with "truth and reality." Once facts are introduced into his account as the basis for certain sentences being true (or false) one is faced with questions about their structure or lack of such—with ontological questions. Of course, one need not ask such questions. Part of the problem stems from some metaphysicians insisting that one has to ask and answer such questions in order to "understand" reality. On the other hand, the opponent of metaphysics insists that nothing is asked or answered. The dispute revolves around the sense of 'understanding.' It is analogous to some ordinary uses of 'understand' in the sense that one is said to understand something about an object when one knows what its parts are and how they are related. But the parts or elements a metaphysician speaks of are not parts in any ordinary sense—as a leg is part of a chair or, even, as an electron is part of a chair. The metaphysician, aside from using analogies, can only make clear what he means by such terms as 'part,' 'understand,' etc., by getting one to see what sort of a question he is raising. So long as his opponent refuses to acknowledge such questions they talk past each other.

We must, in all this, separate the process of getting one to see how a metaphysician uses his terms from the question of what these terms refer to. Or, as some would put it, we must distinguish different senses of 'meaning.' One gets someone to see what a metaphysician is driving at by showing how his terms relate to each other and how different metaphysical positions contrast with each other. To speak of meaning here would be to speak of contextual meaning or, as the early Wittgenstein might have put it, the logical space of metaphysical terms. Yet the metaphysician will also think in terms of the referents of his terms—such referents, spoken of as the meaning of his terms, provide a noncontextual sense of meaning. To guarantee that there are such things that we can speak of "knowingly" some, via a principle of acquaintance, have insisted that one must be acquainted with the elements of one's ontology. Hence, we hear of being acquainted with bare particulars and even with exemplification. This complex of issues we need not go into here. I merely note that, in some sense, the contextual setting of the metaphysician's terms does not exhaust their role or use or meaning. We may note another thing. The critic of Shorter's bent insists that the metaphysician provide a contextual setting for his terms by ordinary illustrations and uses. This he cannot do except by analogy. To be able to do so without resort to analogy would, in effect, require metaphysicians to differ about ordinary facts. This we saw earlier they cannot do. To attempt to do so is to invite defeat, at some point, when one is shown that, after all, one makes use of an analogy which does not, as no analogy can, exactly fit the situation. Be that as it may, we return to the question of the difference between alternative ontologies, since I have advanced the claim that contrasting such alternatives furnishes, in one sense, meaning to each and should serve to get one to see what is involved in a metaphysical question.

Consider the problem of universals. We confront Big Socrates and Big Plato. A philosopher now asks what is it that accounts for the assertions that both are white, what do the two patches have in common? A ready answer is "the color white." But the philosopher wishes to push the question to inquire whether this color white is an element to be included in one's ontological analysis of the two patches. Is there something, in some sense, that they have in common? One might reply that it all depends on what you mean by a "thing." If the color white is a thing then there is *some thing* that they have in common. But is this not just an elaborate way of saying that they both are white? Yet to the philosopher, to say that two things are white and to say what accounts for their both being white is to say different things. So, how does the metaphysician who claims that a universal quality and a relation of exemplification accounts for the identity of colour differ from a philosopher who claims that there are no such things? To deny that there are universals and yet construct an ontology is to say what there is, or what combination of what there is, accounts for the ordinary facts we start from. Hence, a metaphysician may hold, as some have, that the two patches contain particulars of a special kind which account for their both being white—particular whitenesses. In short he introduces (γ), without universals. As it stands (γ) will not do within the terms and context of the metaphysician's question. For it is clear that to speak of particular whitenesses is to invite an account for these things being of the same kind or having something in common. That is, the metaphysician, whatever else he may mean by giving an ontological analysis or account, has clearly not answered his original question by introducing (γ) and particular instances of qualities. His solution invites the re-asking of his question and, hence, is no solution. Irrespective of any objections one may raise to universals as entities, this particular difficulty does not arise on a gambit that acknowledges universals, say (α). To me this shows a difference between the two views. Further, forgetting this difficulty with (γ), one may notice that on (γ) there is only one kind of fundamental entity, instances of qualities; on (α) there are two, universals and bare particulars. This is a further difference. These are the sorts of reasons that lead one to assert an ontology like (α), (β), or (γ) not, as Shorter would grotesquely have it, that Big Socrates is white and square. If one now insists he sees no difference since both say that Big Socrates is white, then we must remember, with Dr. Johnson, that one cannot guarantee to provide understanding.

Shorter objects to a difference I pointed to regarding (α) and (β). He argues that a version of (α) could be stated so that this difference disappears. His objection is instructive if irrelevant. The point was that on (β) no option is open: since a particular must be a combination of qualities, it must have, by, as it were, the "grammar" of 'combination,' more than one property. There is no corresponding necessity on (α). This difference remains. It is, after all, one of the consequences of having simple particulars on one view and no such things on the alternative view. This is why his objection is irrelevant: it remains to be seen why it is instructive.

Suppose, by adjustments in one's ontology or even in the guiding principles one uses in formulating an ontology, two competing ontologies manage to successfully deal, in their own terms, with the criticisms of each other. Do they say anything different? How does one show two formulations are two and not one? Impressed by *the fact* that two "theories" account for exactly the same facts and do not differ, as they cannot, about such facts one may insist they say the same thing. Equally impressed by the contextual differences, including the painstaking changes that dialectical arguments have brought about, both within one ontology's framework and in contrast to its alternatives, one may insist that they say different things about what is ultimately "there." In a sense such a dispute is pointless. The game, as it were, lies in carrying the alternative positions to such a point of impasse and to clearly seeing and specifying what is involved in so doing. In a way one learns more about what an ontology means and involves as one goes along. Thus, the creation of new alternatives, changing the contextual setting, changes, *in one sense* and in that sense alone, the meaning of any one alternative. To insist that the enterprise is hopeless since all alternatives must ultimately say the same thing, is simply to give up the philosophical task. One may do this and repeat, in parrotlike fashion, that all we are talking about is a white square, whether he says so literally or disguises the assertion behind questions and cute examples. But to others this is not an argument, simply a bore. Long ago Bradley held that the denial of metaphysics is itself a metaphysical position. One need neither put it so strongly nor make it so respectable.

www.ingramcontent.com/pod-product-compliance
Lightning Source LLC
Chambersburg PA
CBHW031257290426
44109CB00012B/619